LOST ENDEAVOUR

Charles Watkins, sometime after 27 July 1918, having received his commission in the Royal Air Force.

Charles
WATKINS
LOST
ENDEAVOUR

A SURVIVOR'S ACCOUNT *of the* ILL-FATED
GALLIPOLI CAMPAIGN

*

EDITED *by* MICHAEL CRANE *&* BERNARD DE BROGLIO

LITTLE GULLY PUBLISHING
2023

Text © copyright Charles Watkins

Annotations, biographies and maps © copyright Michael Crane, Bernard de Broglio

Images copyright expired or no known copyright restrictions

All rights reserved. No part of this book may be reproduced in any form by electronic or mechanical means, including information storage and retrieval systems, without permission in writing from the publisher, except by a reviewer who may quote brief passages in a review

This edition first published November 2023

A catalogue record for this book is available from the National Library of Australia

ISBN 978-0-6459276-0-3 (hardback)
ISBN 978-0-6459276-1-0 (paperback)
ISBN 978-0-6459276-2-7 (ebook)

Little Gully Publishing
littlegully.com

CONTENTS

Editors' foreword vii

Foreword 1
Acknowledgement 3
Apologia 5
Preface 7

PART I

Chapter 1	The Splendid Brigadier	15
Chapter 2	People Like Us	23
Chapter 3	Walk the Plank	29
Chapter 4	Moment of Truth	33
Chapter 5	The Baptism	36
Chapter 6	The Guinea Stamp	41
Chapter 7	A Mouse's Whisker	44
Chapter 8	The Long Grass	50
Chapter 9	Over The Top	55
Chapter 10	The Harbinger	68
Chapter 11	A University Degree	73
Chapter 12	The Numbered Bullet	77
Chapter 13	Compensations of Eternity	88

PART II

Chapter 14	The Cross and the Crescent	93
Chapter 15	Strictly Private and Reserved	102
Chapter 16	The Pot Boils	111
Chapter 17	The Philosopher	116
Chapter 18	One Moment in Annihilation's Waste	121
Chapter 19	A Vexed Woman	125
Chapter 20	Money for Jam	127
Chapter 21	My Lady Nicotine	135
Chapter 22	The Cobbers	142

PART III

Chapter 23	Life is A Maze	155
Chapter 24	If I Touch It with My Stick	169
Chapter 25	Mr and Mrs Fly	175
Chapter 26	This Brotherhood Lark	177
Chapter 27	An Innocent Abroad	189
Chapter 28	God Rest Ye Merry, Gentlemen	196
Chapter 29	Run, Rabbit, Run	204
Chapter 30	Exeunt	210

APPENDICES

Appendix I	Charles Watkins biography	215
Appendix II	Notes on individuals mentioned in the text	229
Appendix III	Notes on events and places mentioned in the text	250
Appendix IV	42nd (East Lancashire) Division, Order of Battle and Field State, 2–5 May 1915	258
Appendix V	Excerpt from 'The Lancashire Fighting Territorials'	263
Appendix VI	Timeline, May to December 1915	273
Appendix VII	Establishment, Drafts and Battle Casualties	287
Appendix VIII	Gallipoli Roll of Honour, 1/6th Lancashire Fusiliers	290
Appendix IX	Charles Watkins & 'The Gallipolian'	306

Maps	320
Abbreviations and acronyms	329
Bibliography	332
Editors' acknowledgements	334
Index	335

MAPS

Map 1	Rochdale and environs	321
Map 2	Gallipoli	322
Map 3	Helles	323
Map 4	Morning of 4 June 1915	324
Map 5	Evening of 4 June 1915	325
Map 6	Morning of 7 August 1915 — before the advance	326
Map 7	Afternoon of 7 August 1915 — after the advance	327
Map 8	42nd Division section from 19 August 1915	328

Editors' foreword

Reader, be warned. This book is not a conventional history of the Gallipoli Campaign. These are the personal impressions of a private soldier who fought at Helles with the 1/6th Battalion Lancashire Fusiliers in 1915. That soldier recorded his 'hotch-potch of Gallipoli memories' more than 50 years after the events described. When the editors attempted to align his stories with the historical record *(see appendices II and III)*, it became clear that the passage of time had given rise to a number of factual inaccuracies.

Take the lively tale of the 'cobbers' in Chapter 22 as an example. Charles Watkins relates how two Australian artillerymen turned up at his sniper's post and insisted on 'having a go.' The bag was one Australian killed, the other wounded. The editors tracked down the incident, which was real enough, but found lots of small discrepancies. The Anzacs were said to be 'huge chaps — about 6' 4" and broad with it' but the Australian staff sergeant who was killed stood 5' 8" inches tall, was actually a gunner, named Pearson not Ballantyne, and a recent migrant from Britain to Australia, so probably lacking the pommy-baiting drawl! No doubt Watkins' description of the two Australians is an amalgam of the many Anzacs he met during the war.

Confabulation is again at work in Watkins' description of Christmas Day, 1915. That morning, says Watkins, they woke to find the countryside blanketed by a heavy fall of snow that hid all the ravages of war — 'except for the pink-splashed snow near the trench from the blood of the wounded as they had staggered by after last night's hand-grenade attack.' It is a fact that it did not snow at Gallipoli on 25 December 1915, but an icy blizzard did freeze the peninsula on 28/29 November. Contemporary photographs show the exceptional snow-fall blanketing the ground, just as Watkins described.

Human memory is a complex and fallible cognitive process. When recalling the past, it is not unusual for individuals to form a 'blended memory' from different sources and experiences. This is rarely a deliberate act of deception but rather a consequence of the brain's attempt to create a coherent narrative based on existing fragments of memory. Watkins appears to have merged his memory of Christmas with the November blizzard.

One further example will suffice. In the final chapters, Watkins tells of the Gallipoli evacuation, when he volunteered to man a sector of front-line trench until the last possible minute. Another ripping yarn, wonderfully told. The only problem is that Watkins' battalion had been withdrawn more than a week earlier. Could this be a shared memory, the product of many reunions with old comrades? It seems unlikely that a private soldier, an infantryman to boot, would have been detached for special employment. However, the editors caution the reader against too quick a judgement. Some elements of the 42nd Division did indeed remain on the peninsula until the final night.

So, what is this book and why should you read it?

Gallipoli is not short of operational histories, but few accounts get into the mind of the private soldier so successfully. Watkins described *Lost Endeavour* as 'my own impressions of the Gallipoli campaign' and here the account succeeds brilliantly. The sketches of Gallipoli trench life are well realised, perhaps because they were honed over many nights in the saloon bar. But Watkins was also a perceptive observer, who could relay the thoughts and feelings of his less articulate comrades.

Australian historian John Laffin reckoned Watkins' description of life at Helles to be 'one of the most vivid in Gallipoli literature.' The editor of the *Rochdale Observer*, who was given a copy of the manuscript in 1970, thought it 'one of the most moving pieces of writing about Gallipoli that I have ever read.' And Watkins' stories certainly passed muster with veterans of the campaign, for many extracts were carried in the Gallipoli Association's journal. Its editor thought Watkins' 'compassionate treatment and understanding of the soldier' more than compensated for the book's 'less salubrious episodes' and 'unexpurgated army language'!

Lost Endeavour was privately printed in 1970 for limited circulation. To meet demand, Watkins had a second run printed in 1982. Nevertheless, copies remain scarce, a fact which spurred the compiling of this edition. Watkins deserves to be read by a wider audience.

The editors have faithfully transcribed the original text, retaining the author's idiosyncratic grammar and capitalisation. This includes language used by Watkins that would be unacceptable today. In fairness, Watkins was a product of his age, one when Imperial Britain assumed a cultural superiority over subject nations.

Watkins did not title chapters 23 and 25; these were composed by the editors, who have also added a short biography of Watkins as an appendix, and collected together several articles he wrote for *The Gallipolian*, the journal of the Gallipoli Association.

For background and context, the editors include their notes on individuals and events mentioned in the text, an order of battle for the 42nd (East Lancashire) Division, and detail on the 6th Lancs Fusiliers at Helles — the battalion's establishment, drafts and battle casualties, a timeline for May to December 1915, and a Gallipoli roll of honour.

Michael Crane & Bernard de Broglio

Foreword by
The Viscount Rochdale
OBE, TD, DL

This is an unusual book which I have had the privilege of reading before publication. It is a book which needs to be written, filling in as it does something of the human aspect which must almost inevitably be missing in the more profound war records.

I have, of course, been particularly interested and appreciative for the many kind references to my father — the battalion Commanding Officer — but there are others who appear in these pages, whom as a boy in Rochdale before the War I can remember and whose memory is very rightly placed on record here.

> They shall grow not old, as we that are left grow old:
> Age shall not weary them, nor the years condemn.
> At the going down of the sun and in the morning
> We will remember them.

I commend this book!

(Signed) ROCHDALE
3rd July 1970

Acknowledgement

In the finalising of the original rough draft of this book, I gratefully acknowledge the invaluable assistance of the Viscount Rochdale, OBE, TD, DL, whose sure instinct in these matters prevented me from committing any breach of good taste, and for his help whereby I was able to correct one or two historical errors.

I am especially grateful for the foreword he has so graciously provided for this little book that hardly seems to merit such approbation.

I acknowledge, too, with pleasure, the assistance provided by the following gentlemen:

> Major T. P. Shaw, MBE, of the Lancashire Headquarters of the Royal Regiment of Fusiliers — for the information he provided and so enabling me to include the exploits of the 1st Battalion, the Lancashire Fusiliers.
>
> Captain E. W. Bush, DSO, DSC, RN, who so distinguished himself when serving in this campaign as a very young Midshipman — for his correction of certain Naval details.
>
> Group Captain E. F. Haylock — an old friend and airman turned author and journalist.

To all these my thanks for their valuable assistance and suggestions in the final preparation of this book.

And if I extol so highly the inestimable value of comradeship in this campaign, it is only because I had the good fortune to be serving alongside those best of comrades — the Lancashire Lads.

Apologia

The Dardanelles campaign must surely rank in history as a classic of British military ineptitude, and of British, Colonial and French heroism. I hope this amateurish and oft ill-written personal account of this colourful and tragic campaign might give some idea of the times, and of the campaign itself.

This book makes no pretence of being a historical document, but all recorded incidents are factual — allowing for minor aberrations of memory.

In his Commendatory Foreword, Viscount Rochdale — son of my old Commanding Officer on Gallipoli — has been kind enough to approve from his father's notes, much of what I have said.

I am deeply indebted to him for his painstaking care and check, and for his Foreword to the book, which lends an air of verisimilitude and respectability to my own rather sketchy impressions of the campaign.

Some apology seems to be due too for the occasional and oft bawdy language of this book. But with my regrettable passion for truth, it would be impossible for me to make a cosy Sunday school story of my own traumatic experience on the Gallipoli Peninsula.

December 1982

Preface

In perpetrating this literary outrage, some apology is due. I could give many plausible excuses for recording moments of this disastrous campaign, but the real truth is the selfish pleasure I find in recalling one crowded hour of glorious life. It was my very good fortune to serve with a Lancashire Territorial Division. To the memory of those contumacious, argumentative, sentimental and lovable Lancashire lads — 'Salud.' No better comrades ever trod the field of battle.

This particular campaign has already been much more heavily criticised by far abler pens than mine — criticised by men with greater professional knowledge and training. So — is there any need for yet another spate of reminiscences? It's a reasonable objection. Above all, why seek to resuscitate, at this late hour, memories of a campaign of more than fifty years ago? I can't answer these objections except to note that unrecorded endeavour has so often been the scabby memorial of the British Army, and I myself feel compelled to pay some tribute to my old comrades. History too, seems to impose upon its servants the necessity for some recording of special events, however inadequate such recording may be. And this campaign was rather 'special.' A mere nine months campaign, on a pocket-handkerchief size battleground, and with more than a quarter of a million British, Colonial and French casualties — there should be a story somewhere.

The occasion that sparked off this late splurge of memories was quite accidental. It was during a holiday in Spain last year — the days spent on the sea-shore soaking up the sun and sea, but the magic nights pulling me irresistibly to the little old village of Oropesa about a mile or so inland. There's no place in all the world where humanity so easily lets its hair down and lives so naturally in the easy friendship that God first made for us, as in an old Spanish village at night. A 'men-only' show, with the shy Senoras and Senoritas gathered in gossiping groups on the benches in the Plaza a discreet distance away. But the little tables outside the cafes and bars are full of chattering noisy men, and with the tables and chairs overflowing off the crowded side-walks right into the streets themselves. There was one old man who always seemed to be on his own, with his own little table. Lonely, and completely alone, with his half-empty glass of wine at his elbow. Occasionally a passer-by would call out a greeting to the taciturn motionless figure who seemed to be staring into space. Sometimes he'd acknowledge the greeting, but most times he seemed withdrawn into himself and into — God alone knows — what thoughts. All the little tables packed I sought the only vacant seat I could see, and this was at the old man's table.

'Se puede?' I asked, hesitantly, before seating myself.

He momentarily flickered into life. 'Si, si, Senor,' he replied politely, and clearing a place on the little table for my own bottle and glass. 'Con muchissimo gusto.' Then I invited him to share my bottle, but he demurred, insisting that he himself should act as host. The half-empty glass at his elbow had been much in evidence the last half-hour and I guessed a shortage of pesetas. The need to tread warily was evident, for Spanish feelings and Spanish pride are notoriously touchy. It was then I noticed some faded campaign ribbons above the pocket of an equally faded and rather grubby shirt. I grew bold. In my halting Spanish, and with what I hoped was a friendly smile, I said 'Since when did old soldiers refuse to share a bottle?'

The response was that of a man galvanised into life by an electric needle. His face cleared and his eyes lighted. 'The Senor is an old soldier, too?' he asked — almost incredulously I thought sadly. Not that I blame his initial doubt. When you're turned 70, you don't look like an old soldier... only old.

So bit by bit we progressed. At the second bottle we had words and strong disagreement as to who should bear the cost — this ridiculously low cost of approximately 10½ pence for a litre bottle of good rough wine. It must have been my night, for once again I happened to say the right thing to soothe the too delicate Spanish pride. 'I always heard,' I said, 'that the Spaniards were for ever complacent of the Whims-and-whams of the foreign visitor.' He stared fiercely for a moment, then laughed outright. The laugh transformed him. The ugly scar across his cheek seemed to disappear completely and his dull eyes sparkled. 'The Senor has the silver tongue of an Englishman,' he said smilingly.

'The silver tongue of an Englishman to match the golden heart of a Spaniard,' I replied, and the old man thumped the table in vigorous appreciation. 'Ah, Senor, I drink to you! What a camarado you would have been for me in Morocco.'

From then on, we grew nearer and nearer. For the next two or three hours he regaled me with reminiscences of his soldiering days in Spanish Morocco. Fortunately for me he spoke slowly, with the hesitant slow speech of a man quite old, and piecing together with difficulty memories of a colourful and oft violent past of his days with the Spanish Legion. I was able to understand most of what he said, thanks to his slow speech, for my own knowledge of Spanish wouldn't win me many prizes.

The heat of the desert by day, its unbelievable cold at night. The harsh discipline of the Spanish Legion, the savage fighting with the Moors, the forced marches, the tongue-cleaving thirst, the raw hunger for even an old crust of bread — the even more raw hunger for a woman — all these things came to life as the old man rambled on — relieved occasionally by chuckling reminiscences of riotous dissipation in towns and villages. Wrapped in his stories he hardly noticed the arrival of the third bottle of wine. At last he stopped. 'But I am boring you, Senor,' he said apologetically. 'A thousand pardons, Senor!'

'On the contrary,' I told him. 'You've enjoyed telling me these last two hours of your soldiering days. But not half so much as I've enjoyed listening to you, believe me. Ah! What a story it would make if only I could write it down.'

'Thank you, Senor. But a story not worth the half of a peseta,' said the old man, deprecatingly. 'But you yourself, Senor, you must have your own stories to tell to me — yes?'

'Plenty of stories, but not enough Spanish, unfortunately,' I told him. 'But someday I might get round to telling my own story, even if only to myself.'

'Yes, Senor, you should, you should.'

'Old soldiers, they live again in their memories, and memories are made of this.'

'Sometimes I think,' the old man went on, 'that we old soldiers have a duty to tell to those yet unborn, of the times in which we lived. All too often what we have done, or tried to do, is so easily forgotten — maybe never heard of. And if we don't talk of these things to those who know nothing of them, how will they ever know?'

* * *

Ay! Maybe the old man was right. The trouble is that most old soldiers develop a reluctance to talk — except perhaps after a few drinks, and when we seem, then, to get a little boastful and silly. At best, and when we are stone-sober, we feel we are merely a little boring to a new and unsympathetic generation.

So we clam-up. We leave it to the cold, clinical dissection of historians to record the battles, the victories… and the defeats. The live and vivid experiences of the soldiers themselves are seldom, if ever, recorded — which is a pity, for without these how can the atmosphere of the times themselves ever be made to come to life. And the worst offenders are the British, with their built-in inhibitions, their national gift for understatement, their inherited taboos, their English — oh! so damned English — apologetic modesty. So history is the poorer thereby.

A pox upon such modesty! In their long war-torn history, no nation has encountered more, endured more, and overcome more, than the British.

The forgotten dead sleep on, and those of us who have managed to survive — well, most of us — are not very clever in communicating. To another old soldier — a few reminiscent words, a few grunts of approval, of appreciation — and we find it easier to talk and to know that we are being understood. But to those strangers to the things of war, we feel we are talking in a foreign tongue. So except amongst ourselves, we clam-up. To an old soldier the unforgiveable crime to himself is to appear boastful.

For what it is worth, I'll record my own impressions of this Dardanelles tragedy in a few reminiscences. A little sympathy on the reader's part for my lack of skill in this recording, a little exercise of his imagination — and the reader may get some idea of the times and of the campaign itself from these sketchy reminiscences. That's the best I can hope for.

Students of military strategy and tactics had best throw this book away for they'll learn nothing from it. In fact, I know even less of strategy and tactics than did the High-Ups who conducted the campaign. What's more, a lowly private soldier sees very little of the larger picture of war — his own grubby little nose is always buried too deeply in his own particular patch of the dung-heap.

Allowing for minor aberrations of memory — mainly of names and exact identities, all incidents related are factual. They dare not be otherwise — lest some of my old comrades who must still be left alive would make it their business to hobble along and clobber me with their crutches. As regards names and exact identities, except for the names of Whitfield, Greenwood, Griffiths, Stansfield, MacLean, Lord Rochdale and, of course, the names of the six VCs of the 1st Battalion, Lancashire Fusiliers, any other names mentioned are fictitious. They are fictitious for the very simple and commonplace reason that although I remember vividly the characters themselves, I have forgotten — after this long ago 50 odd years, their exact names.

The flights of fancy occasionally encountered are, regrettably, my own — the occasional flights of fancy of a queer guy who, oddly enough, saw more than he ought to have seen of the romantic side of war. In the last analysis all wars add up to a gigantic futility. But if I dare misquote Sartre, maybe wars have been necessary for the preservation of the spirit of man, as whores have been necessary for the preservation of good women. What interests me most, however, is the human side of the drama and the persistence under all conditions of man's indomitable spirit and of his undeniable quirks.

Just after the close of this tragic campaign and the final evacuation of the armies that fought there, I saw in some American or Canadian newspaper a simple tribute. It was a sketch of an imaginary monument erected on the tip of the Gallipoli Peninsula — a carved figure in stone of a British Tommy — and underneath the caption 'One thing that will never be evacuated — the memory of the British Private Soldier.' After that, it seems there's nothing much more to be said.

I myself was fortunate to survive — right up to the end of this campaign and the final evacuation — and without any crippling wounds or sickness. For this I thank whatever gods there be, and mainly because I was able to stumble across a secret hoard of treasure. I was able to plunge my arms into it — right up to the elbows. It's probably the greatest treasure a man can ever find, the rich golden friendship of men — this rich golden metal, refined to the N-th degree in the crucible of war.

We had caroused together, roistered together, and wenched together; hungered and thirsted together; soldiered and suffered together.

Such things make friends. Such things forge even stronger the golden fetters that bind men to one another.

* * *

> When I remember all
> The many friends, so linked together,
> I've seen around me fall,
> Like leaves in wintry weather;
> I feel like one
> Who treads alone
> Some banquet hall deserted,
> Whose lights are fled,
> Whose garlands dead,
> And all but he departed.

Bedford 1968

LOST ENDEAVOUR

PART I

(A hotch-potch of Gallipoli memories)

CHAPTER 1

The Splendid Brigadier

There's no denying it — he was a splendid figure of a man.[1] Everything about him gleamed and shone — from the peak of his cap, heavily splashed with the 'scrambled egg' of rank, right down to the toes of his highly polished boots and leggings. His square-toed boots fitted snugly in the gleaming stirrup irons. Even the accoutrements of his mount, a magnificent chestnut mare — all the jingling harness, the snaffle chains, the bit, the reins and the saddle — everything shone and sparkled in the morning sunlight. The restless mare tossed her disdainful head as she faced us, giving the Brigadier a chance to display his superb horsemanship. The creak of the saddle and the jingle of the bit and snaffle chains sounded musically on the air. Oh, yes! We were properly impressed — as indeed we were meant to be.

300 years of breeding and family sat in that saddle — a long heritage of good schools, tradition and impeccable breeding looking down on us all, gathered there in sullen silence in the gully. Even the mare seemed to share the same expression of extreme distaste as she was constantly made to face us. She champed her foam-flecked bit continuously and snorted, and her hooves struck sparks from the stones as she fidgeted and pranced beneath him. Like a fractious two-year old at the starting tapes, dead reluctant to face us was that mare.

And in truth, you could hardly blame the mare. I doubt if in all her loyal service she had ever been called upon to face such a motley crew as we.

---

[1] Brig. Gen. Herbert Cockayne Frith, CB. See Appendix II, Notes on individuals mentioned in the text.

Our uniform was now more reminiscent of Mexican bandits than British soldiers. Our headgear ranged from the pith helmets we had landed with; forage-caps — filched from dead artillery men; woollen Balaclava helmets — the gifts of industrious knitters back home; and the slouch and feathered hats of Australian troops, many of these hats snatched from bodies hardly yet cold. These Aussie hats were the most popular headgear of the lot with us — hats that spoke of a life more untrammelled and free in another part of the world. I guess, too, these slouch hats satisfied in some part a not-too-far-back boyhood yen for the cowboy outfit. I dunno! All I do know is that we prized these Aussie hats — at times we could hardly wait for another Aussie to get knocked off so we could pinch his hat.

So, if our motley collection of uniforms and 'hats assorted' upset the mare, you could hardly blame her. I doubt if Captain Morgan had ever commanded such a bunch of sinister looking pirates. All that was missing was the skull and crossbones floating above us.

While we were stood at ease, waiting for the Brigadier to begin his spiel, he would lean down occasionally from the saddle to speak a bit with the Brigade Major and the other two — three High-Ups standing around him in respectful silence. My eye roved the scene appreciatively — well — I always was a bit of a nut for the dramatic scenes of life — one of those queer guys, like when they're kids, the sight of a circus sends 'em sky-high... if you know what I mean. If I could sling words together properly, maybe I'd have made a good war correspondent or something. I wouldn't ever see the suffering, the ineptitudes, the glorious 'balls-up' of all war — all that a chap like me would ever see would be the pictorial glamour of it. Well! — I guess we're all as we're made, God help us!

Yeah! I know I'm a bit of a nut. So what! But it really was a colourful scene — a bit like a stage tableau and with Kipling himself as stage director — and even if the presence of a horse in this ragged and ravine-ridden terrain seemed about as appropriate as a whore at a christening.

When I'd met the mare the same morning about two or three hours before this parade, I'd been on the scrounge down the gully for a possible unattached tin of bully-beef — or maybe a spare Army biscuit or two; always ravenously hungry in those days I seemed to be — never could get half enough to fill the big hole in my belly. But when you're young, active, living a hard life in the fresh air all the time, and only getting just enough rations to support life — well, even cannibalism seems to have its points.

So that's how I came to meet this same mare being led gingerly at walking pace along the boulder-strewn gully from the beach some two miles back. The groom was taking it dead slow, fearful of his charge breaking a fetlock or something. A bit surprised, I asked the groom what-for all this, and got a blast of obscenity that scorched even my hardened ears.

Now God be praised for the blessed gift of cuss words — those ever-present Gadarene swine by which the front-line soldier is able to dispossess himself of all his many devils of anger and frustration. Without the safety valve of the cuss word, the soldier would explode in spontaneous combustion.

An old pal of mine from our training days in Cairo — this groom, and I fell into step alongside him, listening gleefully to him venting his feelings in rich Irish eloquence, and with a line of blasphemy that must have ruined for ever his chances of a harp on the clouds. He cursed both the mare and the Brigadier equally. The curses on the mare were charged with all the affection and emotion of the true horse lover, and with the mare responding occasionally by nuzzling the groom's ear. The curses reserved for the Brigadier — well, all I can say is that if everything came true that the groom was wishing him, he'd be in one hell of a mess.

But for his four-legged charge, the groom was as tenderly solicitous as a mother with a sick child. If there's one thing that moves me, it is to see the close bond of communion between horse and man — 'surpassing the love of women' as the psalm says somewhere. And this bloke surely loved horses. Back in Cairo I'd occasionally accompany him in the early morning when he was out exercising her and a spare horse — 'ride and drive' in the usual military fashion. I'd met him one early morning with these two horses, and I asked him if I could accompany him sometime and ride the spare horse during exercise. And me, of all people! I'd never put a leg across a horse in my life. He seemed a bit surprised, but agreed — probably glad of the company. After a struggle I managed to clamber on to its bare back. After a three-mile walk and trot I got back feeling as if a thousand needles had been stuck in my liver and wondering ruefully if my chances of future fatherhood had not been ruined for ever. Dammit! I never could properly master the trot. But next day he had it fixed up with a proper saddle, and many an early morning jaunt we had along the side roads and paths of the cultivated parts of the Nile, and with the morning mists at times completely enveloping us until the hot sun eventually dispersed them. Ah! Great and glorious days those early morning rides of some three or four months back. As I padded alongside him in the gully that morning it seemed like a thousand years back.

And now this same mare was facing us, snorting disdainfully, in this pre-arranged stage setting, and with the Brigadier like a shining god, mounted and aloof, looking down on us from his Olympian heights — looking down on the scruffy, crummy shambling crew, in serried and scratching ranks arrayed before him.

And by contrast, the Brigadier was — oh! so sparkling and clean, and the shiny coat of his mount gleamed like silk. And I'll bet, too, that even the Brigadier's shirt and vest were just as spotless and clean — not like ours, crummy and filthy. But when you've been wearing the same sweat-stained shirt for a couple of months or so, it's bound to get a bit crummy.

And I'll bet, too, this Brigadier didn't have to do his thrice-daily de-lousing act like us, stripping off your shirt, raking out from under the seams the eggs of these crawling things from Hell that plagued us night and day, cracking between your finger nails their foul little bodies till your fingers were stained with their putrescence. You could never get rid of 'em. 'These buggers are grandmothers in 24 hours', one of our chaps said. I think he understated.

The setting for the scene was perfect. A natural alcove that led off from the gully — an alcove of some 100 yards or so roughly circular, high-walled on three sides by the cliffs, and where a man — or a Brigadier — could soldier on for a thousand years with never a chance of being hit by shell fire or machine gun bullet. And three or four natural little caves there too, tucked snugly in the sides of this alcove, about three feet from ground level, and where the winter rains couldn't reach. A perfect nature-made Brigade Headquarters, and within it this shining Brigadier, martially mounted, and all set to tear us off a strip.

His batman told us the Brigadier even had a real bed, with clean sheets and all. And that some of the caves had bits of carpet on the floor. But we ourselves couldn't expect bits of carpet on our own sodden trench floors, any more than we could expect a bed. It wouldn't make sense.

Under his Brigadier's cap, his face, as yet scarcely etched by lines from his long service, was a deep olive. His tunic of the correct Pukka-Sahib shade, was illumined by the coloured ribbons splashed across his chest — ribbons that spoke of many other campaigns in the service of our far-flung empire. As he faced us, all aloof, god-like and stern, it was pure Kipling.

In marked contrast to us, too, his physical condition was superb, but he didn't keep that way on a haphazard daily ration of three Army biscuits and an occasional tin of bully-beef like we did. We know this, from the occasional fatigue parties we had, carting up from the beach the special rations to Brigade HQ, and it wasn't just biscuits and bully. But you couldn't maintain that god-like physical condition on biscuits and bully, could you?

Against the pale cream background of the cliff walls, horse and rider were silhouetted sharply — a virginally clean background of cliff face, innocent of all the stains and muckment of war — except for a thin ragged ribbon of stain, dark brown, that ran from the top to nearly the bottom of the cliff face. A chap I knew told me all about this stain and how it happened. 'Funniest thing you ever saw in yer life, mate,' he said. 'There was this bloke, see, of the RE Signals, laying a field telephone wire along the top edge of the cliff to go down to the Brigadier's HQ down there, when 'plop' — a sniper's bullet caught him in the jugular vein. Gawd! I can still see it. Fair give me a turn it did — so sudden like.' He stopped, clearing his throat, before proceeding, of a large globlet of phlegm — as if for ever wishing to clear himself of the larger phlegm of war.

'Blood spouted out of his neck like a fountain, it did, mate, and ran all down the cliff side. Then he sorta sank down on his knees and lay flat, poor bugger, and with his face hanging over the top of the cliff.' Then, and as if in memoriam, he interjected, 'Poor bastard.'

'Lay there for a couple of days, he did, mate, and wi' the Brigadier playing bloody hell because the stretcher-bearers couldn't get near the body to shift it — what with all the bloody shelling and that being so fierce,' — and adding sourly, 'Reckon it must 'ave spoiled the old Brigadier's supper that night.'

But apart from the stain it was a nice enough alcove — the sort of place where you'd like to picnic in peacetime.

Then someone called us to attention and the Brigadier started doing his stuff. 'An ignoble army of fed-ups,' he called us, amongst other things. It's true we'd been grousing a bit more than the allowable British Tommy's grouse. It wasn't so much the rations, though God knows they were sketchy enough. It wasn't even the absence of tobacco and cigarettes. We'd got over that one all right when one of our chaps discovered that if you rolled the thick pieces of brown paper the cartridge cartons were wrapped in, it made quite a passable smoke. A bit sweetish to the taste, but still a smoke. It was only after so many of our chaps had gone down sick with D.A.H. (disorderly action of the heart) through smoking these damned home-made Havanas, that someone 'up top' bestirred themselves and the Navy landed us a once-only emergency issue of their beloved 'Tickler' brand tobacco.

It wasn't only these things — it wasn't even the endless casualties. After all, in war you expect to get killed — it's all part of the game.

It wasn't only these things — it was, oh! just everything. A general feeling that nobody cared a damn, a feeling that this was a war being run by bloody amateurs just for kicks, and with nothing achieved and nothing ever likely to be bloody well achieved. Always the same stale old pattern — a bit of ground gained here, a bit lost there, and always the bill to settle with another score or so of casualties. All this, and lousy rations just enough to support life, no smokes, and a dead certain feeling that everything was just a glorious balls-up. An army waiting for Godot. There's a sure bush-telegraphy among troops on active service that is more reliable than all the official bulletins — not that we ever had an official bulletin.

So, if we had groused a bit more than tradition allows the British Tommy, you couldn't blame us overmuch.

Even our officers were gloomy and disillusioned, trying hard, like the good chaps they were, not to vent their frustrations on the men. And I reckon they must have felt the privations even more than we did, what with missing their Officers Mess comforts and all.

'An ignoble army of fed-ups' he called us, among other things, and lashed us unmercifully with his tongue as we stood there sullen and resentful at the injustice of it all. Occasionally during this harangue, an officer would bark out, 'Stand still in the ranks' to the itching, scratching, motley crew lined up there, and then to stoop himself for a long scratch, quite unconscious of his action.

'This is war. Do you hear me, men? This is war', etc, etc. His mare would give another impatient prance and he'd quiet her and resume, 'Behave like soldiers — British soldiers. Do you hear me? Let me hear no more of this. This is war.'

One of our blokes, goaded beyond endurance, said aloud, 'Does he reckon we think it's a f— Band of Hope picnic?' and got his name booked for talking on parade.

With a final prance of his mare and a haughty, 'Dismiss the men,' we returned to our sectors in the trenches.

There's no doubt the Brigadier had really upset us. Maybe, too, in our bitterness we misjudged him. In fact I am perfectly sure we did, because he wouldn't have risen to that rank if he hadn't been a damned good soldier, and although not of our generation, he probably was sure he was doing his proper duty as a good officer in bawling us out like that.

It is a truism that every man is stuck in his own generation. This Brigadier was bogged down in *his* — a generation of stern soldierly duty. What's more, those ribbons across his chest weren't won at Officers Mess tea-parties. We, in turn, hadn't yet had sufficient time to extricate our own feet from the morass of our own civilian background. And you can't convert overnight a 'fish-and-chip' brigade of civilian soldiers like us into hard-baked 'old sweats.' And it couldn't have been easy for the Brigadier himself, used to commanding Regular soldiers only, to have to cope with — and probably for the first time in his soldierly career — the problem of commanding a brigade of 'Saturday afternoon' soldiers (the nickname for the Territorials in Lancashire). And the problem for him, too, was made more difficult in this unique theatre of war, which strained almost to breaking point the endurance of even the professional Regular soldiers. If we'd understood him more, likely we wouldn't have been so damned peeved when he bawled us out. If he himself had appreciated how really raw was the raw material he had to lick into shape, maybe he might have... but what the hell; what's the use of all these might-have-beens. I am merely recording how touchy and proud are these Lancashire tykes, and how we felt after being dressed down like this, however unjust our resentment may appear to the — er — Gentlement of the Establishment.

Back in the trench there was a strained silence — a resentment, black as Hell, at what the Brigadier had said. After a while one of our Sergeants asked me, 'Ere, Cobber, you've bin to school. What does ignoble *really* mean?'

It's true I was a bit more erudite than they. Most of 'em had started work half-time in the cotton mills at the age of 11 or 12, and then gone on to full time at 13. I myself was a bit more lucky. In spite of the prevailing poverty of those days of a non-welfare state, I'd been able to stay at school until I was 15, thanks solely to the incredible efforts of my mother.

I didn't know how to answer the Sergeant properly. And the blokes gathered round him had that shut-in look of bitterness you see sometimes in soldiers after they've been bawled out unjustly after doing a good job of work. I didn't want to offend them — stout chaps that they were, but in a spasm of anger I translated cruelly.

'He means we're nothing but a shower of no-good bloody bastards.'

The Sergeant was quiet for a time, then he let out a long whistle — either of admiration, incredulity or exasperation. 'Us,' he said. 'Us, of all people. I can't believe it.'

But it was the remark of Jock McPherson, the Scots laddie amongst us, who restored our self-respect and set us all grinning like mad. 'Och!' he said. 'What does a f— Brigadier know about real war?' But for quite a time after I could hear the Sergeant muttering unbelievingly, 'Us, of all people. People like us.'

CHAPTER 2

People Like Us

From the mighty armada of ships that carried us to our appointed task, there had been such a burst of cheering taken up from ship to ship, that must have been heard at the very gates of Heaven. This armada of troopships, packed to bursting point with the male stamen of the flower of the British Army drawn from all parts of the Empire. Little brown men of the Gurkha Regiment, slant-eyed, tough as steel. From their belts hung the long deadly kukris — the wide bladed weapon that could cut you in two with one swipe. We used to play football with these chaps back in Cairo — us in our football boots and they barefooted. With their bare toes only they could lodge a drop-kick as good as any Cup finalist. Legend has it that one of our chaps some weeks later in the campaign saw one of these Gurkhas, Kukri in hand, tenderly reviving back to consciousness a captured Turk and then cutting off his head. 'He feel it better that way.' Terrible little men who could move quiet as cats in the dark. If there was an opening in the enemy barbed wire, they'd slip through quietly in the dark, working silent, deadly havoc with their kukris. Like a bunch of ferrets let loose in a rabbit warren, they'd slash and slay, then return just as quietly back to their own trenches.

There were the contingents from Australia and New Zealand — big, brawny men in their picturesque slouch hats, criminally reckless with their lives, and who feared neither God nor Devil.

Men of the Scottish Highlands, their jealously prized and colourful tartan kilts of their training days in Cairo now exchanged for the drab but more serviceable khaki kilt of battle — men who spoke in harsh gutturals in argument. In repose they were dreamy and with the far-away look in their eyes. Over their trenches at night you'd oft hear the strains of melodious Scottish ballads hanging softly in the air. They weren't new to this sort of thing called war. Their own forefathers had done the same sort of thing before — for hundreds of years before — in many a mist-clouded glen, both in their own bloody clan wars and in fighting the hated Sassenach. Now united with their hereditary foe in this common fight, we could still feel their slight suspicion and contempt.

Men of the Irish regiments, who would just as cheerfully fight for the Ould Country against us but now called upon to fight alongside us, just 'rolled up their sleeves and spat on their hands.' To an Irishman a fight's a fight — what's it matter who you are fighting. Men whose grammar had a strange syntax, and whose voices bore that persuasive lilt that could charm a bird down from the tree.

Men of our own English County Yeoman Associations — their horses now in storage back in Egypt until the end of this present campaign, and feeling mighty uncomfortable, too, in their unaccustomed accoutrement of infantry. The very names of these English Yeomanry Companies evoked nostalgic memories from the mists of our own war-scarred history. Descendants of other bygone yeomen who had fought under long-since forgotten English kings, and drawn their long bows with such deadly accuracy at Crecy and Poitiers. In the present hour of their country's need these same descendants had 'resharpened their pikes' and without questioning the rights or wrongs of it had flocked in their thousands to serve. Soft speaking, gently nurtured men — these Companies of English Yeomanry, who strove manfully not to smile at the broad uncouth Lancashire accent of my own regiment. In the whole of the history of the world a more mixed and colourful army was never crowded on to such a small battlefield as on Gallipoli.

Men of my own (42nd) Lancashire Division — men of the Rochdale Battalion, my own battalion, the 6th Lancashire Fusiliers, men of Rochdale, birthplace of John Bright. John Bright, the bitter opponent of slavery, the wealthy cotton mill owner of my home town, who, to his own great financial detriment so bitterly opposed the practice of employment of child labour in the cotton mills. John Bright, ardent pacifist, hater of wars.

'The Angel of Death is abroad in the land — almost you can hear the beating of his wings.' When he spoke these words to a hushed Houses of Parliament at the time of the Crimean War, I wonder did he have a presentiment of this later campaign in which so many of his future fellow townsmen would bleach their bones. Men of my own battalion — at the start of the war their spare and poor physique, product of a neat diet of fish and chips and long close-confined work in the mills — but now filled out to more robust proportions after a few months of good army food, fresh air and exercise.

Men of the 5th Battalion, the Lancashire Fusiliers — miners, and sons of miners most of 'em, more at home with a pick than a rifle, whose clogs had but recently echoed metallically over the flagstones of Wigan and Bury. Pugnacious types of men these, arrogant too in their pugnacity. 'Give us back our bloody clogs and we'll punce these bloody Turks off the Peninsula.'[2]

Men of the 6th Battalion, the Manchester Regiment — the 'office boys' battalion we nicknamed them. Outstanding in their neatness and prissiness, these chaps — recruited from the offices and warehouses of our Northern Metropolis. But valiant youngsters for all their prissiness. In a corner of the notorious Krithia vineyard one day in early August later that year, dozens of these lads lay packed in neat array — caught in a murderous machine gun cross-fire while charging the enemy front line.[3] In a space not a lot bigger than the average large front room of a house, they lay head to foot in layers of their last sleep. Even in death they were neat and orderly, but packed so tight together that to cross this corner of the vineyard you had to tread their squelching bodies, festering and swollen in the hot sun. I doubt if the average age of these lads exceeded 20 or 21.

> For how can man die better
> Than facing fearful odds
> For the ashes of his fathers
> And the temple of his gods.

---

2 Puncing was a break-time activity for mill workers in Lancashire. Two men wearing clogs would sit opposite each other, then kick each other's shins until one gave up!

3 Battle of the Vineyard — Battle of 6/7 August at Helles. See Appendix III, Notes on events and places mentioned in the text.

If Horatius was right, those boys sure agreed with him unanimously.

And now, some 50 odd years later, I wonder if the 'brass' conscious Mancunians ever give a thought to those lads of theirs and their sacrifices in this mistaken and ill-fated enterprise. I'd like to think so, but I guess we all tend to forget. And it's easier — and more comfortable — to forget than to remember. And we can always buy a poppy on November 11th just to square the account.

Somehow, the efforts of any generation are forgotten almost before History has had time to turn a page.

There were umpteen other regiments — just as good — but you'd need a whole library to list the deeds and exploits of this multi-regiment force that sailed so blithely and confidently to their appointed task.

And what sort of a Lancashire Fusilier would I be if I didn't give pride of place to the 1st Battalion, the Lancashire Fusiliers who landed a few days before we did. '6 VCs before breakfast' has since been, and justifiably so, the Battalion's proud boast, as at 4 a.m. on the morning of the 25th April they successfully stormed 'W' beach from open boats. To quote from General Sir Ian Hamilton's despatch:

> No finer feat of arms has ever been achieved by the British soldier — *or any other soldier* — than the storming of those beaches. Gallantly led by their officers, the Fusiliers literally hurled themselves ashore, and, fired at from right, left and centre, commenced hacking through the wire. A long line of men was at once mown down as by a scythe, but the remainder were not to be denied…

Or to quote from the despatch of Vice-Admiral de Robeck on the naval aspect of the operation:

> It is impossible to exalt too highly the service rendered by the 1st Battalion, the Lancashire Fusiliers, in the storming of the beach. *The dash and gallantry displayed was superb…*

No wonder they were allotted 6 VCs for that early morning's work.

Bromley
Stubbs
Grimshaw
Willis
Richards
Keneally

Their ranks ranged from Major to Private — you'll find 'em all, individually by rank, in the *London Gazette* dated 15.3.1917. But I have omitted their individual ranks, for in sublime heroism like this, as in Death itself, rank has no meaning.

The cost was heavy — 11 officers and 350 men of the Battalion killed or wounded in the first few minutes, besides 63 of the 80 naval ratings manning the cutters and the open boats that towed ashore the boats of the Lancashire Fusiliers. No aircraft cover, no protective landing craft — nearly a thousand men, touched by the fire of heaven, jumping from their open boats and hurling their frail bodies against the stout shore defences — the barbed wire that ran past the water's edge into the shallow sea itself, the hidden machine gun nests, and the murderous rifle and shell fire. And losing more than a third of their number in so doing, but pressing on and attaining their objective — justifiably earning their '6 VCs before breakfast.'

It's never been done before. I doubt if it will ever be done again.

These Lancashire lads! As they say affectionately in the North of England of these Lancashire lads — 'they're a right lot of monkeys.'

But I think you'll agree — good monkeys!

On the bridge of our own ship, looking down at the sea of excited faces in the bow of the ship — the excited boyish faces of our Territorial Battalions keen to emulate the example of the 1st Battalion, stood the sombre, craggy figure of the ship's Captain, distinctive in his dark blue serge jacket against a background of the khaki-clad officers, equally excited and now crowding the bridge. I happened to glance up at him.

> God save thee, Ancient Mariner
> From the fiends that plague thee thus
> Why look'st thou so…

The lines of the poem flashed in my mind as I watched him. I thought — maybe he's got a son of his own somewhere in the world, caught up in this Armageddon. He stood there, tense and absorbed, like a man in a trance, his hands clenched tightly on the rails of the bridge.

But the face that looked down on us was full of infinite sadness and pity — the Aztec father sorrowing for his children on the eve of the sacrificial knife.

CHAPTER 3

Walk the Plank

The voyage from Egypt to Gallipoli had been one full of joyous excitement — marred only by one sad little incident. That was the day a member of our Battalion had been made to 'walk the plank.'

Back in Cairo where we had done our training, we were housed in the barracks of the Citadel — an ancient edifice on high ground just on the edge of the city, an ancient and imposing Moorish type of stronghold of historic proportions, and one that legend says was built in the days of Saladin. The outgoing regiment we had relieved there in 1914 had left behind them a cocky little white-haired fox terrier answering to the name of Jim. We adopted him, or rather did he adopt us. More regimental than the RSM was this little tyke, and just as well versed in regimental routine. So, if he wasn't exactly on the official ration strength, he was just as much a member of the battalion itself as our own CO, Lord Rochdale,[4] and just as popular. Each morning at daybreak as we'd set out for our gruelling training marches over the nearby desert, he'd accompany us from the barrack square as far as the Citadel gates, just ahead of the band. The two great iron-studded doors would be rolled back — necessitating three men to move each door, and giving us a glimpse of the view below of the city of Cairo at dawn. 'Earth has not anything to show more fair' begins Wordsworth as he gazes on London from Westminster Bridge. Ah! — nuts to London, and to Westminster Bridge! I guess the poet had never bumped up against the breath-taking

---

[4] Lieut. Colonel Lord Rochdale (George Kemp). See Appendix II, Notes on individuals mentioned in the text.

beauty of a Mohammedan city at dawn, never seen the delicate tracery of the tall minarets climbing up into a pink-grey sky, never heard the disembodied voices of the Muezzims from their high eyries calling the faithful to prayer, never heard the call bounce from minaret to minaret, as in turn each Muezzim took it up.

This city in the early dawn has a sort of other-world majesty that can hit you like a blow under the heart, and makes you feel sometimes like you wanted to blub. Yeah! I know it all sounds daft, but if you've ever seen Cairo at dawn you'll know what I'm talking about.

As we'd trudge down the steep hill leading from the Citadel to the city below, our heavy ammo boots sending up low clouds of choking white dust, the distant beauty of the city below never failed to silence even the most raucous of us — irreverent lot of buggers that we were by any standards. All you'd hear above the silence would be the shrill barking of the little dog getting fainter and fainter. Every day he'd accompany us just as far as the Citadel gates and wait just outside, barking us a shrill God-speed. At times, the Colonel, Lord Rochdale — ahead of us, would lean low from his saddle and make a playful lash at him with his whip, sending the dog into transports of doggy delight. At dusk on our return, he'd trot out to meet us about a quarter of a mile from home, do a smart about-turn, and keeping about 50 yards ahead of us, and in a dead straight line, he'd head the procession and lead us in, one ear stiffly erect. I doubt if even a line of trees would have deflected him from his straight course as he marched solemnly in front of us. It was as good as a circus act and pleased us all no end, from the Colonel himself right down to the humble Private. And with a battalion of boy soldiers such as us, a dog is almost a sine qua non. We sure loved that little tyke.

And if Jimmy belonged in part to all of us, he belonged by common consent in particular to one Jack Gibney.[5] Jack was the one the little dog always sought out among the crowd of us lounging in the evening on the barrack square, deftly and good-humouredly dodging our good-natured kicks as he roamed amongst us seeking his rightful master. A quiet, rather inarticulate lad was Jack, and not over-given to conversation, but he was the adored god to this little tyke, the one to sit slavering

---

5 Probably Pte 8620 John McGibney, KIA 15 May 1915, age 20. See Appendix II, Notes on individuals mentioned in the text.

at, open-mouthed, as dogs do to a beloved master. Not much of a dog lover myself, I do however recognise that between certain individuals and dogs there is a bond of understanding that is an alien world to me. Jack and this dog were on the same wavelength. Even speech was superfluous between them. So it was particularly hard on Jack that he should have to be told that his pal had to walk the plank.

When we left Egypt, and despite stringent instructions to the contrary, the dog embarked with us for Gallipoli — either smuggled himself along with us, or had been smuggled on. His presence aboard the ship had been tolerated the first few days and the ship's crew vied in providing him with the choicest delicacies. But about a couple of hours from our destination, the ship's captain had decreed that the dog must be shot — and shot within the hour.

I guess the captain hadn't much choice. Probably he considered that the chance barking of a dog over the quiet sea when we anchored for the night might increase the known hazard from enemy submarines. Jack at last mournfully agreed to distract the dog's attention while one of our sergeants — a crack shot — administered the coup de grace.

The kaleidoscope of memory plays weird monkey tricks with us all at times. A shake of the tube, and the little bits of coloured glass at the bottom of the tube throw up a bewildering variety of pictures. Something has jogged my arm right now and I get a vivid picture of us all collected at the stern of the ship to witness the execution — the mournful fascination of the macabre — and drowning our own deep concern in the blessed sea of levity.

'Aim that bloody gun straight, Sarge,' muttered one of our chaps, anxious to get the whole rotten business over quick. It could have been a swift and merciful end for this little tyke. The position was perfect — the dog sitting in his usual stance of open-mouthed, panting adoration of his master.

Only just in time did someone knock the gun up and the bullet whistled harmlessly out to sea, and a very real tragedy was averted by a split second. At the last moment Jack, in an agony of indecision, had thrown himself protectingly around the dog and picked him up.

Language abounds in clichés — and the English language most of all. The one I call to mind is that one about the right moment finding the right man, or something like that. I give full marks to the middle-aged

South African war veteran, one of the three or four in our battalion who, in the stunned silence that followed, crossed over to Jack and consoled him. Then he took the dog in his own arms, patted it a bit, then gently dropped it over the side of the ship.

The fast receding figure of this little white terrier paddling frantically after us evoked our coarse and cruel merriment.

'Come on, Jimmy, you little bugger — swim for it.'

'Swim for it, you little bugger.'

We jeered and cheered at this little morsel of sentient life paddling after us so desperately until he eventually passed out of sight of our straining eyes.

I give up — trying to understand my fellow men. Now I just accept 'em. Did we merely take out our cruelty for an airing, or did we do it to hide our own deep compassion? Is it that the opposites of Life must always go hand-in-hand — cruelty and compassion, good and evil? Or is the secret key to our electric cosmos the necessity for the twin opposites, the positives and the negatives of everything? Without Evil, there'd be no necessity for Good. Maybe without God there'd be no necessity for the Devil. It seems one can't exist without the other.

But I'll tread no farther in these dangerous quicksands, but leave it to the psychologists and psychiatrists to sort out. Me — I have a peasant's mind, and with a peasant's passion for simplicity in my thinking.

After tea we discussed the problematical tomorrow and our landing ashore, while Jack assuaged his outraged feelings by picking a fight with a perfectly innocent soldier of the 7th Battalion who, he said, had been laughing at him.

But Jack needn't have worried all that much. I stumbled across him two or three days later on the Peninsula. He was lying stretched out, his eyes wide open to the noon-day sun. His body had the rigidity of the final sleep, but his face bore the relaxed, happy expression of a man who has at last whistled up his lost dog.

CHAPTER 4

Moment of Truth

By early evening we had arrived within sight of the shores of the Gallipoli Peninsula and the ships cast anchor for the night about a mile or so off shore. We stood along the rails and drank in the vividness of it all: the setting sun sparkling on the calm Mediterranean waters, the occasional flying fish breaking surface and descending with a flash and a plop back into the calm lake around us. The low-hulled destroyers of the Royal Navy weaved snakily around us in endless circles of protection from enemy submarines. Farther off, like big growling watchdogs, the light grey-painted cruisers and battleships loomed protectingly behind us, their big guns trained shorewards. The unnatural silence around us was occasionally shattered as their big guns let off a broadside towards the land with a bang that rattled your very teeth. From the land itself could be heard the faint crackle of rifle fire, and flashes from bursting shells there showed intermittently of the fighting already going on there by the 29th Division who had landed a few days earlier. Our hitherto high spirits became dampened as the evening wore on. Men tended to avert their eyes and our normal blasphemy-laden language reached a new high in purity. Seems like the influence of the Sunday School and its pious platitudes are more far-reaching than we realise! Apprehensively, we young soldiers knew we were fast approaching the Moment of Truth. This Moment of Truth! For young soldiers to encounter it for the first time — it's a solemn experience. There was none of the brave old 'We, who are about to die, salute thee, Caesar' — this brave old Roman salute to death in the arena. Rather was it 'We, who maybe are about to die, are a lot of frightened little boys.'

As always at such times, rumours of already fantastic casualties circulated the ship and we sought our hammocks that night in uneasy sleep.

Throughout the night we could hear the restless mules stowed in the big transport moored alongside us — hear them stamping and shoving as the muleteers moved among them, calming them when the big guns let off a salvo.

Gawd! What blokes they were — these muleteers. The only ones who could do anything with those obstinate long-eared bastards on four legs. Little brown-skinned men from the hills of Northern India — little brown men of frail physique, perpetually frightened eyes but with guts of steel. Without these blokes, many of us would have perished of hunger in the later stages of the campaign, when the only means of getting food and ammunition to us in the front line was by means of pack-laden mules being led along the narrow treacherous paths cut out on the sides of the steep gullies, where down below the winter rain and snow had created a quagmire. Day and night, night and day, through rain, snow, frost and fog, these little brown men led their protesting mules, bringing supplies to us with clockwork regularity. Awaiting anxiously the life-saving food, rum and ammunition, we'd wait until we saw the two or three of them appear in sight, greet them joyfully on arrival with obscene abuse and cusses. They'd grin delightedly, sup gratefully the hot tea we made for them, share our smokes, and in broken English return in full our affectionate curses. But we well understood one another.

'White man speak with double tongue' says the Comanche Indian. Soldiers, too, 'speak with double tongue.' The one you hear is the language of the barrack sewer. The one you don't hear is the language of the heart — the glint of the eye, the warm grin of brotherhood. These muleteers, their teeth a white slash against the background of their dark-mud-covered faces as they laughed merrily at our obscene greeting, and gave us in return, curse for curse, smile for smile.

> I, too, saw God through mud —
> The mud that cracked on cheeks when wretches smiled.
> War brought more glory to their eyes than blood
> And gave their laughs more glee than shakes a child.
> I have perceived much beauty
> In the hoarse oaths that kept our courage straight,
> Heard music in the silentness of duty...

Not many men on the Peninsula did a lousier, more thankless, more unglamorous and more dangerous job than did these muleteers. Their only weapon, the rope that dangles from the mule's head as they led their protesting and vociferous four-legged charges along the slippery, sliding narrow paths cut on the side of the gully. From these deadly dangerous narrow paths both mule and muleteer would occasionally hurl headlong to destruction into the sea of mud below. It's not a job that earns many VCs — in fact I doubt if any of them ever earned a medal.

But in Valhalla, there'll be a special lot of seats at the tables of the mighty for these little brown men of frail physique, with the frightened eyes and the guts of steel.

CHAPTER 5

The Baptism

From the moment we set foot ashore we felt the strangeness and magnitude of the task. Unlike our brothers of the glorious 1st Battalion, we were not battle-trained. The narrow beaches were swept continuously by shrapnel fire, taking a heavy toll of us at first. I will admit we wilted a bit under our first baptism of fire. This was not at all like we had been led to expect in war.

The psychological shock was profound.

Back in Egypt our training had consisted of long gruelling marches in the desert, open-order skirmishing, followed invariably by a fixed-bayonet charge at a non-existent and mythical enemy. None of us had ever heard a shot fired in anger, much less been on the receiving end of a bursting shell. It was like landing a bunch of tough Boy Scouts on a battlefield. The heavy staccato of enemy machine-gun fire didn't sound a bit like the more ladylike cracks of the blank cartridges we had fired with such enthusiasm in our mock skirmishing. But worst of all was the low raking fire of the shrapnel which tore holes in our ranks. There was consequently indescribable confusion, quite a bit of panic, and after the first shock we did what we should have done in the first place — what we should have been told to do before landing — dodged for whatever shelter we could find in the lee of the low cliffs, and crouching behind every stone and blade of grass. Looking back, I am astonished that none of us had been given the least guidance from higher authority of what to expect or what to do. Presumably we were just to land, and the rest was up to the Almighty to sort out. This lack of planning, this lack of direction, became for me all too evident as the campaign developed. Or maybe I didn't see enough of the larger picture. I dunno.

Matters weren't improved by the supporting guns of the ships firing on too low a sight. To be caught between enemy guns and your own is an experience that would unnerve the stoutest soldier. I have a vivid recollection of our Battalion Commander, Lord Rochdale, harassed and shaken, but not more shaken than I was, wondering how to tell the ships to lengthen their range. A rumbustious type of Lord of the old school, this Lord Rochdale, whose full figure and florid complexion was already showing the strains of the Mediterranean heat. I was a lean rangy type of youth myself and the heat didn't bother me none. Eventually I found a battalion signaller and brought him along to our Battalion Commander. I often wonder if this pathetic little 'flag-waggler' ever managed to get his shaky message through to the ships, or was it just coincidence that the ships' guns mercifully lifted their sights soon after.

A great bloke, this Battalion Commander of ours, veteran himself of the South African War where, when he was then a mere Sir George Kemp, and before he became 'His Lordship,' he had commanded a troop of local yeomanry. I was too young to see the return of those yeomanry in those more picturesque war days, but they tell me that as they rode through the town on their return, sunburnt, slouch-hatted, clattering hooves, and sheathed sabres swinging from the saddle, they tell me it was a sight fit to bust your heart with jingoistic fervour. But of the realities of the present campaign, he was as ignorant as the Generals above him in rank. Still, I'll not hear a word said against him. He did his best, and his ignorance of the existing war conditions was no more than many of his superiors.

Whenever I think of this particular Lord, I am always reminded of a meeting my father took me to when I was quite a youngster, just to hear this then Sir George Kemp address it. After his long and well-received speech about something or other, blest if I remember what it was all about — probably something political — anyway it's not important. But what I do remember vividly is the anecdote he told at the end of it to illustrate some point he was making. A story about the interment of some cowboy who had bitten the dust in some saloon gun brawl. His mates were searching their hearts for some virtue or other to write on the rough wooden cross over his grave, as was the custom, Western fashion. But search as they would, they couldn't in all honesty recall one single virtue of the deceased to scrawl on the bit of biscuit-box wood from which the rough cross was fashioned. He was a wife-beater, a drunkard, horse thief, bank robber, rustler — in fact everything a decent cowboy didn't ought to be. But they couldn't send him on his last round-up without some word of praise. Finally they settled on:

He dun his damndest — angels could do no more.

In war, there's all sorts of soldiers; good soldiers, bad soldiers; but most of us twixt and between. You might think of a better epitaph for any one of them — but I'm blest if I can.

Gawd! The way these old soldiers do natter on. At this rate it'll be 100 years before I get this book finished. But I'd be a right surly dog if I let this moment pass without some tribute to a beloved Battalion Commander.

> Breathes there a man with soul so dead
> Who never to himself hath said
> We are the ones he loved and led,
> A few still alive — but most of us dead.

Maybe it is that a Territorial Unit such as ours feels a closer attachment to its Commanding Officer than does a Regular Unit. COs of regular units tend to get posted away to other units. A CO of a Territorial Unit like ours was the same CO we had in peace-time, as well as in war-time. We went abroad with him as a close-knit town unit. Our comrades were the same boys we had played with as kids, grown up with, worked with and scrapped with. So it's natural that we should all feel tightly attached. And the CO himself was, as one of our chaps said, 'one of us, like — a chap from the same town.'

Beloved in war by his gang of boy soldiers. Beloved, too, in the days before the war, by our fathers. They, these older chaps, our dads, loved him mainly because he was all that tradition said a real live lord should be — hail-fellow-well-met, flamboyant, good-hearted, 100% for king, country and the British Empire (and to hell with all wogs). In appearance bucolic, smiling or stern as occasion demanded, a good employer, a healthy 'port-wine' complexion, and with a nice sense of the dramatic — even to the distinctive red or white carnation he always wore in his button hole as he was driven daily to his woollen mills — Kelsall & Kemp — in the centre of the town, in one of those early type noisy motor cars. Hats would be touched respectfully as he passed and if you happened to be the lucky one he acknowledged by an old-fashioned and courteous lifting of his hat — well, you'd be boasting about it all night in your favourite pub.

The hard-baked, hard-swearing, hard drinking numerous navvies would straighten their toilsome backs as he passed, and smile happily as they touched their forelocks. They adored him, these hard-drinking navvies and claimed him as one of their own. 'One of the lads' they dubbed him. Maybe too, they adored him because he was a 'traditional' Lord. Or maybe because he was the reputed author of that scandalous couplet that appeared in our local paper during the height of the town's temperance campaign. Lasted a whole bloody month, that campaign did, and with lurid posters splashed all over the town on the evils of the Demon Drink. At school, as kids, we sang lustily little songs of the Joys of Total Abstinence. I even remember giving my own father a childish lecture on the evils of drink. 'Beer,' I told him, 'makes your liver get all bloated and swell out.' 'And whisky,' I further informed him, 'makes your liver shrivel up and get all knotted!'

'Will I be all right, then, son, if I mix 'em?' he asked me solemnly. But I didn't know the answer to that one. And what's more, I mistrusted that owlish twinkle in his eye.

But it was an uncomfortable month for the hardened drinkers — they tended to slouch in at the back doors of the pubs, instead of swaggering in via their glossy front portals. I think it was that couplet that restored their self-respect.

> Damn his eyes
> Whoever tries
> To rob a poor man
> Of his beer.

'Pure poetry, that,' said my Dad, himself an ardent devotee of the product of the Divine Hop.

And this particular Lord was the bloke who was our Battalion Commander. In battle arrogant and contemptuous of danger, almost as if he thought 'Damn their temerity, thinking that a bullet could hit me — me, of all people!' — but always terribly concerned for the safety of his men.

No wonder we thought ourselves specially privileged. We felt he belonged to *us*, and to nobody else, and that made us very proud.

CHAPTER 6

The Guinea Stamp

> The rank is but the guinea stamp,
> The man's the gold, for all that.

But this particular Lord of ours — our Commanding Officer, the first Lord Rochdale, qualifies under both headings, both for the guinea stamp and for being all of a man. God knows, I am no lickspittle of the aristocracy. I am a staunch and rude democrat; they don't come ruder than me, but for this particular member of our aristocracy and for the many others like him in the two World Wars, I am the biggest lickspittle of them all.

I found myself that first day, and by pure chance, right alongside him as we scrambled up the low cliffs in twos, threes and fours in this broken country. Woefully short of breath and with the sweat streaming down his face, revolver in one hand and walking stick in the other, urging his men for'ard — this Lord Rochdale. By rights, and at his age and physical condition, he should very properly have been occupying a safe seat at the War Office or at some HQ Base, and where he could have been and without the slightest loss of honour, instead of being on this young man's Commando-type enterprise. But not for him the soft seat of war. Like many others of his kind, he just wasn't that sort. And he stuck with us during the campaign, sharing equally our dangers and hardships until age and its infirmities finally caught up with him.

CHAPTER 6

It was about the late autumn time of the year. By now our erstwhile Battalion Commander had attained the rank of Brigadier General, commanding the attenuated 125 and 127 Brigades of our own 42nd Division. With his spirit still willing but his flesh weak, he made his undignified exit from us towards the end of the year, on to the hospital ship. Sorrowfully we watched him being hoisted like a sack of potatoes, on to a horse's back. Wracked with rheumatism, lumbago and the kindred ills of middle age, in addition to some other disability — legacy from his former service in the South African War — he had the greatest difficulty in clinging on to the horse being led at walking pace to the beach where a Naval pinnace was waiting to take him to the hospital ship.

He deserved a more glorious exit than that. But it's one of life's little ironies that all too often an undignified exit awaits many a worthy man — be it a felon's cross or the indignity of being carted away like a sack of spuds on the back of a horse.

'Ee! But Ah feel right sad, now 'e's leaving us,' said one of our chaps.

'And 'ow does ta think *he* feels?' said our Platoon Sergeant compassionately.

Indeed, he was very ill; so much so that the hospital ship had to put him ashore at Malta if he were to survive at all. For a time it was touch and go with him but fortunately he pulled through, although for some time afterwards when he first got home he was on crutches and sticks.

Hats off to a very noble Lord. And to a very gallant soldier. I didn't meet him again until after the war. Except for looking very much war-battered, he was back at his duties in his mills again after the war. And, dammit, he even still wore that daily fresh carnation in his button-hole — just as flamboyantly as ever. Goddam this old aristocracy, but you can't help liking these noble Lords of ours. They seemed to stand for something so solid and permanent, something so — so damned English... if y'know what I mean. Arrogant, snooty sometimes, just a little stupid perhaps sometimes — jovial, flamboyant, contemptuous of all other nations — those 'lesser breeds without the law' — but game to the very last, these old time members of the aristocracy. Even peevish and envious democrats like me, we find we can't help liking 'em. And even their very faults make them so incredibly and endearingly British.

And there must be something, too, in this 'hereditary' business, especially with these noble Lords. 'Like father, like son', as they say, for in the Second World War, damned if Lord Rochdale's own son, the present Viscount, didn't go and take up where his own dad had left off, repeating the pattern — he, too, serving with distinction as a Brigadier.

My town of Rochdale has every reason to be proud of these two most illustrious sons.

('And now,' says my Recording Angel, 'stow your blethering loquacity and get on with the story.')

Sorry! — but a chap gets carried away sometimes.

CHAPTER 7

A Mouse's Whisker

After about an hour or so, some of us managed to edge round into a steep gully. The shelter here was very little better and a new dimension of terror was added to our unaccustomed ears as the noise of the bursting shells echoed and reverberated off the walls of the gully. I crouched there trying to will myself to the size of a flea and wishing my legs hadn't been so long and sticking out so far behind the sheltering boulder. In spite of my fright, I was fascinated by the dreadful beauty of the bursting shrapnel that searched the gully so unceasingly. These low bursting shells above us—as they burst they emitted puffs of white smoke that lazily spread out like opening white flowers patterned against the background of the blue Mediterranean sky. It was pretty to watch and made you tend to forget the spitting death their appearance heralded.

There's lots of eerie dreadful beauty on a battlefield—beauty, and music, too, of a kind. Where, tell me, in all the world is there a more pleasing sound than the musical cuckoo-like cough of a battery of French '75' guns, firing in rotation—from the French sector on our right. To hear them on a spring morning on Gallipoli, as was I often to hear them, was to remind me of bare-footed walks as kids back home, treading the dew on the long grass, and with the sound of the ever-present but ever-unseen cuckoo telling us that 'Summer is icumen in.'

The Supreme Artist seems ever to manifest Himself—even among the horrors of the battlefield.

While I crouched there I was wishing fervently I'd never yearned for military glory. Like Falstaff, I'd have swapped the lot for a pot of ale and safety. But as I lay there I was heartened by watching as fine a piece of Irish bravado as ever a man could wish to see. It nearly gave me a heart attack, watching those swaggering silly born bastards, but at times like these it's a tonic to see a bit of this stuff now and again.

Two middle-aged 'old Sweats' of an Irish regiment — judging by the green cloth flash on the side of their pith helmets — were on the skyline, evidently repairing a broken field telephone wire. The bursting shells around them were as thick as flies round an open jam pot but by a miracle they survived. After a few minutes and with their job completed, they stood up, stretched, and straightened their backs, deliberating for a couple of minutes, while down below in the comparative shelter of the gully, we held our breath as we watched them. The shells were so thick around them you'd swear not even a fly could live there. Then one of them and with maddening slowness, abstracted his 'bacca tin from his pocket, filled his pipe, then passed it over to his mate for a fill. After they'd lighted their pipes and got them drawing to their satisfaction, they slung their rifles over their shoulders and with a touch of swagger typically Irish, strolled back to where they came from.

You stupid swaggering Irishmen — it wasn't fair to be giving us heart attacks like you did as we watched you.

But this gallant example served me in good stead. Often caught out myself in similar circumstances during the campaign, I'd remember these two Irish blokes, straighten my back, put on a bit of a swagger — but always somehow managing to end up bolting like a frightened rabbit for the nearest funk-hole!

I'll bet the patron saints of those two swaggering Irishmen were working overtime that day. What's more, being Irish, I'll bet those two chaps had 'RC' stamped on their identity discs.

Probably the most important part of a soldier's kit, these identity discs. There used to be two of them, about the size of a penny, one red, one green, and made out of fibre board. These identity discs were worn on a piece of string around the neck, next to the skin. I forget what used to be stamped on the green one — it could have been first and second choice of flowers in a certain eventuality. But I honestly forget what was on the green one.

On the red one, however, was stamped your name, initials, regiment, and the initials of your own particular fancy in the religious line, like CE (Church of England); RC (Roman Catholic); OD (Other Denominations, or more tersely abbreviated by us to 'Odds and Sods'). Besides being decorative — and until you've seen a bunch of soldiers coming in from a bathe dressed in nothing but their identity discs — well, you've missed a lot, but besides being decorative they had a real practical value and were no end helpful to the Celestial Traffic Cop, busy in those early days directing the stream of traffic crowding towards St Peter's Gate. The main road would be full of the more numerous CEs, and with the RCs and ODs being allowed to filter through now and again from the side roads. So you see, to go into battle without your identity discs would be inviting real trouble — like going abroad without a passport, or walking through the Italian quarter of a town shouting 'Shoot the Pope.'

Being RC meant that we were of the Only True Faith (God forgive us for our presumption), and also ensured — or so we hoped — that we could pass through those Golden Gates without too much passport formalities. As regards the others — well, no doubt there'd be all sorts of difficulties — visas, three references, and all that. So you see, there were all sorts of advantages in having your identity discs stamped RC.

RC meant, too, that among other things we believed in fairies, and the power of our protecting saints to shield us from all harm. And if the protecting saint fell down on his job now and then, and the bullet ripped into your guts instead of bouncing harmlessly off you — well, even a saint can slip up sometimes. So, like I said, I'll bet the protecting saints of those two swaggerers from the Emerald Isle were working frantically that day.

I turned to comment on this to the Lance Corporal crouched alongside me, but he didn't reply. He'd just had half his lower jaw shot away with a clump of shrapnel balls. For a few brief seconds his eyes held mine in the mute agony of the badly wounded. I remember, and still with disgust, the sickening feeling of elation I got when I saw it was he and not I that had been hit. It is only very recently I have learned that such reaction as was mine is a normality in human behaviour, but until I knew that, the sense of shame I first got had haunted me for many, many years. Why doesn't someone tell us about these things earlier? That little black devil of shame you keep trying to push down deep into the well of forgetfulness: it forever keeps bobbing up, and it is only after the thing is explained to you properly that you feel like a man suddenly released from a dark dungeon.

As I saw the torment of this Lance Corporal alongside me, I toyed with the idea of ending his sufferings with a merciful bullet, but the presence of the other chaps not too far away bothered me a bit. Or maybe I wasn't sufficiently adult to make the big decision. And it's not easy for a young soldier in his first day of action to abrogate unto himself the power to dispense final relief by a merciful bullet. Some days afterwards we collected, amongst many others, his identity discs from his unprotesting body, thus formally registering him as having been killed. I hope he died quickly.

I left this tormented chap lying alongside me and dodged to better cover and forgot the whole incident for quite a time. It got tucked away into forgetfulness, along with the endless pile of things to which we all bore unwilling witness. Nature is merciful to us in many ways, but never more so than when she gives us the capacity to forget many of life's bitter moments.

There was no let-up in the shelling. The only one now left in my immediate vicinity was my Platoon Officer. To advance further was manifestly impossible, and to remain where we were was merely to shove yourself to the front of the queue for the Golden Gates. So by unspoken mutual consent we dodged back — he to the deeper shade of the gully and me inside a tiny cave nearby. In the gloom of this little cave were two soldiers of my own gang. One was sat up, his rifle pointed towards his foot. In his eyes as the glare of a man temporarily adrift from sanity. His companion was squatting alongside him, more relaxed, and with an expectant air.

When I entered their cave the voices stopped, and I got that sort of welcome a man gives to his mother-in-law when she comes for a long stay. I enquired if anyone was hurt. In a tense voice, the chap with the rifle pointed towards his foot and said: 'I'm going to get on that bloody hospital ship if it's the last thing I do.' He intended to put a bullet through his foot and limp down to the beach.

Youth is seldom censorious — it tends rather to accept. So I discussed with them the ways and means of it — all of us quite oblivious of the military heinousness of the self-inflicted wound. His companion suggested that a blanket rolled thick round the boot would erase any tell-tale marks of the shot. Then I raised the practical point that getting the boot off afterwards would be a bit messy and painful. The would-be aspirant for the hospital ship flashed me a look of hate. Shortly after, the shelling eased temporarily and we split up and left the cave and with the experiment uncompleted.

Three months later his companion was killed — shot cleanly through the heart. The other chap, so bent on making a hole in his foot, later got himself decorated for a bit of real bravery — humping on his back a wounded comrade from No-Man's Land, and with the air so thick with bullets you'd think it was a bloody hailstorm. Which only goes to show, and as the Lord Buddha has taught us, that the distance dividing a coward from a hero is no thicker than the thousandth part of the width of a mouse's whisker.

I didn't meet up again with this chap until just after the war, by which time I had left my regiment and got myself a commission in the Royal Flying Corps, and with whom I served for the last two-and-a-half years of the war. It made a nice change — the comparative safety of the clouds and enough to eat every day, and being able to wash and keep clean, and I enjoyed it no end. I was on leave when I met him and he was by then demobbed and delivering a load of coal to my house. Surprise was mutual, and after we'd both agreed it was indeed a small world, what else could we do in self-defence except repair to the 'local' to celebrate. He was a bit diffident at first about accompanying me — 'wot with me, wi' me dirty 'ands and face, and thee, all poshed up like' — and it was only after I'd threatened to clout him if he didn't talk sense that he laughed and made his horse comfy with a nose bag. Horse and cart left standing quite safely, we settled for the pub round the corner.

The beer was a bit rough at first, but later, when the publican recognised me as the son of a man who must have been a founder-member of his pub, he produced some beer worth drinking—and this stuff wasn't too plentiful just after the war, believe me.

And we sat there and reminisced. And how we reminisced! This 'under-the-counter' beer was strong stuff and loosened our tongues freely. I studiously avoided any reference to the cave incident but he brought it up himself. With a self-conscious laugh he said: 'Do you know, Charlie, I could 'ave kicked yer guts in, stopping me getting on that hospital ship like you did.'

I protested, 'But I didn't stop you. You stopped yourself. Anyway, you've done all right for yourself since—what with that decoration you got and all.'

He was quiet for a time. He cleared his throat and a stream of coal-dust coloured spittle stained the white sawdust of the bar-room floor. 'Ay,' he said, 'I s'pose I did. But there's things I saw later I could have done better without seeing. Things that leave bloody scars on yer memory—and you find yerself wishin' you'd never seen 'em.'

The first day's fighting had been enough for me. I was cold, hungry and tired. Above all I was tired. The Mediterranean night began to fall quickly and I looked round for somewhere to sleep. Nearby I espied the slouch hat of an Australian, and with its owner lying there with his overcoat as blanket. I crept under the coat to share the warmth. Within seconds the nightmare of the day was blanked out in the merciful oblivion of sleep.

> Oh! Sleep, it is a gentle thing,
> Beloved from Pole to Pole,
> To Mary, queen, the praise be given
> She sent the gentle sleep from Heaven
> That slid into my soul.

Times there are when I think the poet himself must at one time have been a soldier.

CHAPTER 8

The Long Grass

At dawn next day I awoke. I felt a real reluctance to face another day — like I feel a reluctance to get on with this yarn. I haven't the skill, the know-how, the command of words, much less the mental equipment, to paint the picture properly.

The early morning was bitterly cold. So was my bedfellow of the previous night, strangely cold. But if you're minus half a skull you do get a bit chilly. Poor devil! He must have been killed the day before and all unknowingly I had been sharing the couch of a corpse. It says much for the resilience of youth that it affected me very little. I peeled off his overcoat. It was a beautiful coat of the light coloured barathea khaki cloth of the Australian Officer and after I'd removed the insignia of rank, I tried it on. A perfect fit, too. Except for a few bloodstains on the collar it was just the greatcoat I needed and I wore it right through the rest of the campaign — incidentally thereby earning the nickname of 'Cobber' with my mates, a nickname that stuck to me ever after with my Company. My own greatcoat had been discarded, along with the rolled blanket and full pack, soon after we had landed. In common with most others we had divested ourselves of this crippling burden on our backs, for very survival's sake.

Oh! The insane worship by the Old Army of the obscene god of the Full Pack. I often wonder how many hundreds, maybe thousands, of needless casualties had been caused in the early stages of the '14 War by the crippling, penalising immobility of the full pack. The night before we had landed we had been issued with a blanket to be carried on top of the full pack, 80 rounds of ammunition per man to be slung round us — all this on top of what we were already carrying — and that was plenty, believe me. If you think all this is nothing much to hump, try walking just across the street, on your back the valise containing heavy greatcoat, rolled blanket and rubber groundsheet slung on top. Then 80 rounds of ammunition pouched around you, haversack with all the bits and pieces of a soldier's domestic life, cutlery, soap and towel, spare shirt and socks, mess tin, full water bottle, entrenching tool and handle — in fact everything except the kitchen sink. Try this as an experiment and wearing heavy army boots. And if it still doesn't bother you none, try running in it on a hot summer's day, climbing steep ground and with all hell after you. So, although we were to regret it later, what else could we do but ditch the lot, retaining only our rifle, ammunition and water bottle.

As we settled down to trench warfare later, it didn't take us long to collect similar bits and pieces again from off the many chaps around who wouldn't need them anymore and who had retained various odd items. Most of us sought to replenish our deficiencies in kit from the Australians. Not only was their stuff of better quality, but occasionally you'd find around their unprotesting waists the ever-popular 'blow-belt' of the soldier, that so useful hollow leather belt, and which, if taken from the body of an Australian soldier, was usually stuffed with gold sovereigns. It didn't do to be too finicky about the smell for some of these chaps had been rotting in the sun for weeks. From some of our more regimentally-minded officers and NCOs there were dark hints of the penalties of looting, but this left us cold. How could it be looting if even the crows plucking their bones so cleanly had discarded the other stuff as worthless? And always we'd salve our consciences with the tag 'they can't take it with them.'

So I cherished this greatcoat I had acquired. In the pockets I discovered two army biscuits and, joy of joys, some crumpled cigarettes and a small bottle of rum.

There's nothing like Active Service for developing the high sensualities, whether it be the longing for a woman or the more mundane things of life like eating and drinking. These biscuits, the like of which I had hitherto gulped down with such distaste, now possess a thousand and one subtle flavours. Such is the magic of hunger — that indescribable anatomical magic that imparts to the taste buds all the joys of Creation itself. And the rum — oh, boy!

> Rum that can with Logic Absolute
> The two and seventy jarring Sects confute,
> The subtle alchemist that in a trice
> Life's leaden metal into gold transmute.

Not half a dozen miles away from where I now live in Bedford, dreams away its day the little village of Bletsoe. In the famous little pub there is a small plaque commemorating the particular bar of this self-same pub, where Edmund Fitzgerald — no doubt inspired by the self-same sort of rum — conjured out of the air his jewelled words as he translated the Persian poet for our all-time joy. And if I've slightly misquoted thee, Mr. Fitzgerald, for my own purpose, I crave thy pardon. Many others must have unwittingly done the same, leaning up against the bar counter, and struggling to recall the lines in glassy-eyed and sentimental poetic binge.

As I drained the bottle of its last little drop, my own leaden day became streaked with gold. By now the whole Turkish Army wasn't big enough for me to tackle single-handed. And the fags, good Virginian tobacco at its best, and it's at its best when you're craving for it. Sure I am that if God made the world in six days, on the seventh day did He make rum and 'baccy.

Later in the morning I located some of my own battalion and was collared by the Machine Gun Sergeant to try and get our own machine gun into action. This machine gun! If there's been no excuse for my profanity hitherto, I could justifiably now speak of it with every other word a cuss-word.

Among the manuals of military training, hidden somewhere in the archives of the War Office, was a secret document that says every fresh war must begin with the training and with the equipment of the previous war. Not that I've ever actually seen this document but it simply must have existed. How else could we have been saddled with this monstrous piece of old-time armoury?

A relic of the Matabele campaign, this particular machine gun. With its fixed-position tripod when mounted standing a good three feet high, so high in fact that a bicycle-type saddle was incorporated on the tripod for the gunner to sit, it was a wonderful museum piece. During our machine gun training in Cairo, our Machine Gun Officer had told us solemnly that 'in Matabeleland the grass grows very high and it is necessary to have the high tripod to fire over the long grass.' He also passed on to us lots of other priceless scraps of information culled from his own experience of other colonial wars, such as the danger of drinking too much whisky before sundown. With whisky at 6d a tot, and with our pay of 1/- a day, we found the information stood us in good stead!

As our machine gun training progressed beyond the 'crank handle on buffer firing, pull belt to left front' elementary catechism of the procedure for firing to the more sophisticated points of machine-gun handling and usage, I raised the point of this high tripod seat and questioned its use in modern warfare. But I was told sternly that a soldier's duty is to obey orders and not to question the wisdom of higher authority. My pertness did not pay off. 'What's more,' said the Machine Gun Officer, 'since you're so smart, you can stay and clean the classroom after parade.'

Ah, well!

A Mark II Vickers water-cooled machine gun which, with its outer casing full of water weighed about 80 lbs — just the gun alone without the tripod. I know so well the weight of this bloody gun: as No. 2 of the team it was my duty to have to carry the damned thing on my shoulder, on top of all the other personal equipment of a soldier, as we moved from place to place on the Peninsula. We nicknamed it the Tea Kettle. The name was not inappropriate. It had a little screw tap at the end of the gun barrel water-casing, and near the muzzle, and where the steam could escape when the water started boiling when the gun was in action. On a cold morning and long after the gun had stopped firing, the spout of steam on a cold morning was a marvellous give-away of our position. I sure learned to love that little gun — even more than the Devil loves Holy Water.

We managed to get the damned thing up the cliff and once on the top, triumphantly erected it on its tripod. A real sitting duck of a target. All that was missing was the Aunt Sally sign 'Three shies a penny.' Almost the first shell put it out of action and wounding one of our chaps. The tripod itself was badly bent and put out of immediate use but the gun itself — bad cess to it — was unharmed and lived to give me many more delightful days in which to enlarge my vocabulary of cuss words. We dispersed in wild disorder and carried on the rest of the day with the scattered pockets of fighting near the ridge of the gully.

The grass grows long in Matabeleland. It should have grown quite long, too, on Gallipoli just after the war, nurtured and manured by the bones of those lads so often needlessly sacrificed by similar stupidities, on the altars Ignorance, Apathy and Indifference to the realities of this later war.

Just before nightfall the remainder of our Company was collected and assembled into some sort of order and we relieved a company of a Highland regiment in a section of a newly-won trench where they had been badly mauled.

CHAPTER 9

Over The Top

Next day I awakened in the unfamiliar surroundings of a trench — the first trench I had ever occupied. This one so newly-won from the Turks was well constructed except for its firing step facing the wrong way. An improvised firing step had been constructed from sand-bags, facing the enemy. In the gathering light of the day I stuck my head over the parapet and took a long look at the ground ahead, and at the blokes lying there.

From the comparative and comforting closure of a trench it's a sobering experience, this first long look at death in the open. It's true I'd seen many such corpses lying around this last two days, but events then had moved swiftly — everything was on the move and in open country. It was then a bit like a nightmare moving picture, but moving so fast that it had the comforting air of unreality. But this view from the trench was no unreal moving picture. This picture was 'for keeps.' It was a ghastly 'still' that bit deep down into your consciousness. You felt that men were never meant to die like that — frightened and alone — their limbs twisted into such grotesque patterns as they had striven so desperately to loosen the silken cord that bound them to their mortal agonies. Some of this Highland regiment lay sprawled on all fours, killed as they were facing the enemy, their bare arses cocked defiantly at the sky — cocking a last derisive snoot at the Reaper before he finally gathered them in. Some lay peacefully as if overcome by sudden fatigue. The horrid smell of death was everywhere.

Many artists have tried to portray No-Man's-Land — that baleful Sports Ground of the Fiends. None succeeds. At best you only get a faint idea of it. It's a place of ever-changing moods.

At night, and with the barbed wire glistening in the moonlight, it can appear almost benign, tempting you to take a walk and explore. When the searching glare of the green Verey lights would illumine the inky black of the sky it bore a slightly science fiction air.

But at dawn it's at its most sinister. No artist, no paint, no canvas can catch the utter desolation of the idle flapping in the dawn wind of the tunic of the corpse leaning grotesquely on the barbed wire. Like a monstrous Guy Fawkes doll he gibbers obscenely to the rising sun, his teeth bared in the fixed and idiotic grin of death. Only a battlefield corpse knows how to grin so idiotically at the Great Sport of War.

His mates, lying alongside him in the dew-wet grass wait in vain for the rising sun to warn their ever cold bones.

And no artist can ever re-capture the smell of decay, the frightening stench of corruption borne towards us on the morning breeze.

But in spite of all this, this bloody place of Gallipoli fascinated you. There was a 'feel' about this place that was sort of special. The glory, the pathos and the shit of war — and the feeling that what we were now doing in this Homeric contest had been done many times before in this particular part of the world — the feeling that we were just another generation of actors re-enacting an old familiar scene that had often been played before by earlier generations of actors on this same familiar stage. I'm bloody sure this place is haunted. Dammit — you could almost hear the clank of sword upon shield.

I once commented on this point of view to a mate of mine, but he seemed only concerned for my sanity. 'Oh! It's orlright, tha knows, mate,' he said soothingly. 'Tha's just a little bit screwy, that's all. 'Ere, 'ave a little sup o' this tea, an' tha'll feel better.'

Blessed are the insensitive, for they shall ne-er be troubled! But I still say that if you listened you could hear the clank of sword upon shield, and almost feel these blokes crowding round you like you sometimes feel, if you're a Catholic, the crowding of Angels at Mass.

Talking of miracles, our Company Cook produced out of the thin air a real breakfast — bits of fried bacon to help down the Army biscuits, and even hot sweet tea. It was just like being at home again. We sat on the fire step and munched our breakfast and agreed that this was the life. Later, at about 10 a.m. we were told there was going to be a special issue of rum. Oh, I say! This is really spoiling us!

'Beware of the Greeks bearing gifts.' I forget who it was who said that, but it had some point. We were too unsophisticated in the practice of war to know what this presaged. But by the end of the day we had learned our lesson good.

A generous dollop of rum was issued to each man and we were told we were to go 'over the top' at 10.30. This shook us a bit but once outside an issue of rum nothing seems quite so bad to the British Tommy. He needs it at such times. Peace loving by nature, it takes a drop of stuff like that to wind him up — something to transform him — even if only temporarily, from a decent ordinary human being into a savage unthinking animal who sees in the vaingloriousness of war a certain Divine justification. The miracle of War — the only thing that can revert decent human beings back in Time by thousands of years, into the elemental savagery of pre-historic man. Well, that's war, I guess, but don't ask me of all people to explain the greatest puzzle of human progress. I only played my own small part in this International Amateur Dramatic Society, and my interest was not in the ethics of the drama, merely in the manner we ham actors played our respective parts.

Cpl I——, an 'old sweat', with the ribbons of the South African War and other campaigns,[6] finished his rum in one gulp. I took a sip of mine, but not wanting it much after such a good breakfast and happening to catch sight of him looking at me toying with it, I passed it over to him saying I didn't want the bloody stuff anyway. His gratitude was profound, out of all proportion. Twice he thanked me fervently. 'Just what I wanted,' he said. The second time he thanked me was just before we scrambled over the top of the trench. By nightfall he didn't have to bother any more about rum — not, of course, unless there's such a thing as Celestial Rum. Along with Sergt R——,[7] he was killed that night, trying to bring in from No-Man's-Land one of our blokes who'd been hit and was lying there groaning.

6 See Appendix II, Notes on individuals mentioned in the text.
7 Ibid.

CHAPTER 9

Poor Cpl I——; if I wasn't such a born lazy devil, I could give him two or three chapters all to himself. And we could have been such good friends, too, but for that horrid cloud of dark suspicion that always dogged our relationship. And all because of his damned teeth.

The only one in our battalion with a set of false teeth. Before going into action he always took them out and carried them in his tunic pocket — 'so they doesn't get broke', he said. I never could quite follow the reasoning of it, but there it is. Ex-regular of some Cavalry Regiment, veteran of many a skirmish of the Khyber Pass District and the South African War, in addition to his pay he enjoyed a long-service pension of 21 years regular service in the Army, and which was religiously exchanged in the Wet Canteen every night in Cairo for a skinful of beer. 'Couldn't sleep, I couldn't, mates, unless I gets my eight pints every night.' But with good beer in those days only 1½d a pint, his pension was more than enough to keep his kidneys well flushed.

His regular service finished, and like a fish out of water away from army life, he had joined our Territorial Unit in 1913 and was a source of wonder and awe to us boys. Tales of the North West Frontier would fire our young imagination and longing for military glory. He was — par excellence — all that we imagined a soldier should be — tough, brawny and a hard drinker. Hand-carved by Kipling, this bloke.

In addition, he was the most tattooed man I'd ever seen. Specimens of the tattooists' art in India adorned most of his body. One brawny arm sported a carved stone cross mounted on a rock against which the waves dashed in vain, and the design bore the caption 'Rock of Ages.' The other forearm had a more sentimental ring — a flowered circular wreath inscribed 'Mother.' Conventional tattoo designs these, found on the arm of many an old-time soldier. Besides being a virility-plus sign with the old sweats, tattooing was a legendary shield against the risk of 'clap.' But the 'piece-de-resistance', the one that never failed to delight us was the one that covered his body. If you stood near him when he was taking a shower you'd be treated to the sight — and in 'glorious technicolour' — of a hunting scene, the horses on his left breast galloping over his shoulder and following a pack of hounds that trailed down his back, chasing the elusive fox that the tattooist had portrayed just disappearing up his rectum, and all that was visible of the fox was its tail at the base of the spine. 'Three months that tattoo took me, mates, and me going every night till it was finished — and I don't remember how many hundreds of rupees it cost me, too.' A colourful bit of work by any standards and one that left us goggle-eyed in admiration. Fortunately, Cpl I—— had never married; one shudders to think of the effect of this particular tattoo on a girl's wedding night.

There had been some medical hesitation just before we left England as to whether Cpl I—— should be allowed to sail with us, on account of his false teeth. In those days they were almost unknown in the Services, but he had pleaded hard and eventually won the day. 'Can't desert the lads, now can I, Sir, wot with me with my experience and all that.' But the Medical Officer still hummed and hahhed. 'And what happens, Corporal, if your teeth get broken — there's no Dental Lab. in the Field, y'know, Corporal.' But Cpl I—— scored the winning point. He cocks a respectful, humorous eye at the MO and asks, 'But we don't 'ave to eat 'em as well as kill 'em, do we, Sir?' And the MO laughs, folds up his stethoscope and says, 'All right, Corporal, get your shirt on and get out of my sight. But don't blame me if anything happens.' So the Corporal sailed with us, teeth, fox and hounds, 'Rock of Ages,' 'Mother,' the lot.

I was one of his most devoted admirers; his tales of soldiering on the NW Frontier held me spellbound. There was something about him, too; an elemental simplicity that appealed to me. And I found it very gratifying, his massive belief in the superiority, the invincibility, the 'everything that's best' assertion of the British Army. To Cpl I—— there were only two sorts of soldiers in the world — the British Army and the rest, the 'bloody wogs,' that contemptuous band of heathens who occasionally were so misguided as to oppose us. Mine was the devotion seen in that picture 'The Boyhood of Raleigh', the wondering romantic youngster swallowing wholesale the yarns of the old campaigner, yarns redolent of the mystique of the Indian Bazaar and Kipling's India. In return he was patronising and affable to me. We could have been such close and life-long friends but for those bloody teeth.

It was as hot as Hell as we sailed through the Med that night in August '14 — all hatches battened down for the night and all of us lying sweating in our hammocks in the inky black below deck, all lights dowsed because of the risk of enemy submarines. I had a thirst that dried my mouth like a furnace. What's more, I'd forgotten to fill my water bottle before dark. Still, I remembered there was a tap a few yards away. After much groping I found a tin mug on one of the dining tables underneath where our hammocks were slung. The dirty water, cold tea or whatever it was at the bottom of the mug I slung out through one of the port-holes. There must have been an old spoon or something at the bottom of the mug, for something metallic rattled as I slung its contents away. I managed to find the tap, slaked my thirst and slept soundly till breakfast time. Round the mess tables there was a hell of a row going on. Cpl I——, breathing fire and slaughter, was searching for his lost teeth 'wot I put to soak last night in a bloody mug on this 'ere table and now they're gorn.' Angrily he interrogated every one of us, gnashing his non-existent teeth in his fury, and all the young soldiers really scared and trying hard not to laugh. But woe is me, cursed from birth with a perverted sense of humour, I got a vision of the top and bottom sets clacking aimlessly in Davy Jones's locker and frightening the poor fish to death. So my protests of innocence were unavailing. A muscular fore-arm shot out and 'Rock of Ages' hung menacingly over my head. But he was restrained from violence by a nearby Sergeant and when the matter was safe enough to be a general joke I was able to relieve my feelings. However, he got a new set in Cairo some two or three weeks later.

Sorry, Cpl I——, I didn't dare tell you before. But now you're in a land where teeth are not a prime necessity, maybe you'll understand it was all an accident. All the same, but for those bloody teeth of yours we could have been such good friends. And you could have taught me such a lot about soldiering instead of leaving me to find it out the hard way.

* * *

And to think of it, all this needn't have spoiled our friendship. Maybe if I'd told him then what I'm telling him now, he'd have laughed it off. And maybe he'd have still been safe in England when the war was over, but for wilful disobedience of a whole lot of orders. And 'Disobeying an Order whilst on Active Service' looks twice as bad on a charge sheet.

There had been a rigorous medical sort-out at Cairo when we got ready for Gallipoli, and Cpl I—— was one of the half-dozen or so rejected—not because of his teeth this time, but because of varicose veins. When we lined up on the Citadel barrack square, ready to march off to entrain for Alexandria and thence to embark, Cpl I—— should have been parading with the other medical rejects destined to be shipped back to England. Instead, he boldly took his place, fully kitted, in our ranks. A stern Regimental Sergeant Major, himself a much-beribboned old campaigner, dressed him down good and proper as only an RSM can. 'By the time I've inspected the rest of the Battalion, Cpl I——, and if you're still here, you'll be on a charge. And you know what that means, don't you, Corporal?'

'Yessir.'

'Right! Don't let me see you here when I got back.'

'No, sir.'

Maybe the RSM forgot to come back. Or maybe 'old sweats' like these two had a species of double talk that was incomprehensible to youngsters like us. But Cpl I—— sailed with us.

For his 'wilful disobedience of an order, whilst on Active Service' let's hope Cpl I—— got tried and sentenced in a Higher Court, and with St Peter himself as Commanding Officer.

'Prisoner and escort, 't'chun. Quick march. Halt. Left turn.' And St Peter himself, dead stern, and with his Adjutant sat next to him, looks at the charge sheet in front of him and reads out:

'Cpl I——, you are charged with "While on Active Service on the morning of April 27th 1915, when ordered by the Medical Officer to parade with the sick, lame and lazy—to wit, the medically unfit, you did absent yourself from the said parade, and paraded instead with the battalion assembled for entraining for an 'unknown destination.' And further, that when ordered by the Regimental Sergeant Major to leave the Battalion and join the medically unfit parade, you did wilfully disobey such order." Have you anything to say, Cpl I——?'

'Well, Sir, it was like this 'ere. I was…'

But the Celestial RSM shuts him up and barks out, 'Silence, Cpl I——. Stand to attention and stop shuffling yer feet.' (RSMs always act like this—maybe just to remind the culprit he hasn't got a hope in hell of getting off the hooks.)

'Well, Sir, like I was a'saying. I was with the lads, see, wot I'd always bin with,' etc., etc., and he trails off into a long and unconvincing defence of his wilful defiance of an order.

And the CO, St Peter, ruffles through Cpl I——'s Form 252, that tell-tale historic document of misdemeanours that is kept for every soldier. And the 252 for Cpl I—— could have filled a big book, a long and inglorious record of the average soldier's petty offences. Only Cpl I——'s 252 was bigger than average on account of his longer service. Things like 'Being found in a place marked Out of Bounds in Calcutta — to wit, a brothel'; 'Drunk and disorderly' (17 of these); 'Being in improper possession of a bottle of whisky, property of the Officer's Mess, Delhi'; 'Assaulting a native policeman in Cairo'; 'When reprimanded by Cpl Stringer of the Military Police, Cairo, for being slovenly dressed, did use foul language in reply, telling Cpl Stringer to 'go and f— himself' (perform an anatomically impossible feat). The list was endless and St Peter conferred gravely with the Adjutant alongside him. Then he asks the RSM 'And how is it that Cpl I—— comes before me today, Sergeant Major?' And the RSM answers, 'Disobeying another order yesterday, Sir. An order given by Sergeant R——. To wit, when told by Sergeant R—— "not to be a fool and go out there, but to wait until the shelling's eased up a bit" was insolent to the said Sergeant R—— and said, "He's me mate, Sarge. I'm going out there to bring 'im in and y'can do wot the bloody 'ell you like about it, see?" And then, Sir,' went on the RSM, 'Cpl I—— goes out to the back of the trench and tries to bring in Private —— wot 'ad bin 'it in the groin wi' a bit o' shrapnel, Sir.'

Then Sgt R—— is called in as witness. These military trials are nothing if not fair and the accused is always allowed to hear the evidence and to question it if necessary. So Sgt R—— contributes his little bit.

'Sir.'

'It was like this 'ere. We'd just taken that bit o' trench yesterday and got it all secure like, when come nightfall we 'ears a groaning out in the open. Some of our lads 'ad bin 'it over the top, and this one wot we 'ears was about 50 yards away, and we could 'ear 'im quite plain. So Cpl I—— says it was one of 'is mates and as 'ow he was going to bring 'im in. But just then the bloody Turks started shelling thick and 'eavy.'

'Watch your language, Sergeant,' said the Adjutant, sternly.

'Begging y'pardon, Sir. A bad habit wot I picked up from the lower ranks. But when I thinks of them heathen bastards killing good British lads, it makes me that mad.'

The Adjutant smiles a bit in his moustache and murmurs, 'A quid pro quo, surely, Sergeant?'

'As you say, Sir,' says the mystified Sergeant. 'Well, Sir, the bloody — beg pardon, Sir — the bleeding Turks started shelling thick and heavy, and I says to Cpl I——, I says, "Best wait till this bloody — this bleeding shelling eases up a bit before you goes out there," but Cpl I—— says, 'e says...'

'Well, Sergeant, what did he say?'

'Well, Sir, I 'ardly likes to repeat it. But 'e says, 'e says, "Bollocks to you, Sarge. That's me mate out there, Sarge, and I'm goin' to fetch im in, see? And you can do wot the bloody 'ell you like about it." Them's his actual words, Sir, 'is actual words,' concluded Sgt R——, and the Adjutant hides a smile under his hand and St Peter finds himself with a troublesome cough and gets a little red in the face.

And Sgt R—— continued, 'So 'e goes out to the back of the trench and I goes after 'im, just to remind 'im he's disobeying orders and to tell 'im 'e's under arrest. And mebbe to 'elp 'im a bit, in case 'e gets stuck, like. Well, y'see, Sir, Cpl I—— and me was old boozing chums, if y'see wot I mean, Sir. Many's the nights out we'd 'ad together with the bints in Cairo.'

In a voice so sharply edged it could have cut through steel, the Adjutant says, 'Kindly confine yourself, Sergeant, to the relevant details. The Commanding Officer is not interested in your deplorable private habits. What happened next?'

'Well, Sir, I don't rightly know, but then I finds myself afore you, giving evidence, like.'

'These things have happened before, Sergeant. In fact they've been happening quite frequently lately from your part of the world. You've been posted to a New Unit now, Sergeant, but you'll soon get used to it. What's more, we'll soon straighten you out. That's all, Sergeant.'

Then St Peter confers a little with his Adjutant before weighing off Cpl I——.

'Cpl I—— , you have been charged with "Whilst on Active Service you did, on the morning of April 27th, 1915, etc., etc., and furthermore on May 6th, 1915, on Gallipoli, when given an order by Sergt R——, did disobey such order and said, etc. etc." Do you elect to receive my punishment Cpl I——, or do you elect to go for trial by General Court Martial?'

'No, Sir. I'll elect to take your punishment, if it's all the same to you, sir.'

Then St Peter passes judgment. 'Very well, Cpl I——. You'll be admonished and confined to this place for the rest of your days,' and then he whispered something to the Adjutant about that paragraph in King's Regulations and Army Council Instructions that mitigates such offences, mumbling something about that para that says something about greater love hath no man than to lay down his life for his friend… or something like that. Then the Celestial RSM bawls out 'Prisoner and escort, attention! Right turn! Quick march! Left, Right, Left Right! Pick them feet up! Halt! Dismiss!'

Then, as is the way of all good RSMs the world over, he had a friendly word with the culprit, advising him in future to 'watch it.'

'Keep out of bloody trouble, here, Cpl, and y'll be all right,' and then sends him off to his new quarters and calls out after him, 'Look in at the Wet Canteen, Cpl, for a pint o' beer, and tell 'em to chalk it up to me. And, oh! There's another thing — y'll find yer mate waiting there for you, too. And tell him, as well, to have a pint on me.'

* * *

After the prisoner and escort had been marched out, the CO says, 'Well, Adj., shall we go to the Mess for a drink?' But the Adjutant was drumming his finger tips on the desk in front of him and seemed not to have heard. 'I've been thinking, Sir…'

'Yes, Adj.?'

'Well, Sir,' and with a wave of his arm he indicates our little terrestrial ball spinning down there in space, 'if you'll pardon a real earthy expression, they've got a right lot of really rum buggers down there, don't you think, Sir?'

The CO smiled. 'I do, indeed. And if you'll pardon a most unheavenly expression, Adj., we've got some equally rum buggers up here, too. And now, are you coming to the Mess for that drink?'

* * *

'Cowards die many times before their death — the valiant only taste of death but once' says Shakespeare, and with me particularly in mind. He's right, mind you, but he needn't have been so damned blatant about it. In any case, I maintain it was more a vivid imagination than cowardice.

Just before scrambling over the top, Cpl I—— button-holed me solemnly, saying, 'Now look 'ere, Charlie, when you goes over the top, forget all about that dashing forrard at top speed like we did when we was training in Egypt. And wot does these silly buggers know about soldiering, anyway. If yer dashes forrard at top speed y'are just as likely to run into a bloody bullet as if y've taken yer time, see. Just keep a steady walking pace along o' me. Then yer won't be all buggered up and out o' breath when you gets to the enemy front line. 'Cos if yer name and number's on the bullet, it'll find yer, and if it ain't on, well y'll be all right. Oh! And thanks for the rum, mate.'

So I walked alongside him smartly; he like he was going on some distasteful parade and grumbling at the heat, and carrying on a cross-fire talk with the shells bursting around us, in the mordant wit of the old campaigner.

'Missed me that time, didn't yer, y'bastard. Next time, maybe.'

'Gawd! That bugger was a bit near mate, wasn't it.'

But by gosh, how I sweated. I was hit a thousand times before I reached the enemy trench. A high explosive cut off my left leg at the knee, I had my head blown off (twice) and besides feeling machine gun bullets tear into my testicles. I tell you, it's hell having an imagination like mine.

'Many times before their death' says Shakespeare. I beat him there. Mine must have been a million times, I reckon.

I couldn't believe my luck when I reached the wire entanglements unscathed. They had been well pounded by the ships' guns. We crossed them — some of us anyway — and I jumped down into the enemy trench. Fortunately it was unoccupied, for during the bombardment the Turks had retired along their communication trench to their rear trench. I started to do what we had been told to do when we got there — shifting the sand bags from one side of the trench to the top of the other side, and had my rifle laid alongside me. The head of the communication trench leading off to the rear was only a few yards away from me, and I was shocked to see a Turk suddenly appear from around the corner of the communication trench, and with his rifle pointing at me. Blind panic sometimes makes us act instinctively right. My hand encountered the bag of emergency rations we had been issued with that morning, and carried slung on the hook of my tunic, and I flung it at him, deflecting his aim. By now I had my own rifle ready and was about to return the compliment.

I guess fellows like me will never make successful warriors. He was a youngster of about my own age and the look of appeal as I drew a bead on him still haunts me sometimes. And it kept my finger idle on the trigger, but an officer of our battalion and with the best of intentions and approaching him from behind, shot him. I thanked the officer, but without enthusiasm.

'Nearly had you that time, lad, didn't he?'

'Yes, Sir. A pity he had…'

'Had what?'

'To get killed like that, Sir.'

There was a puzzled look in the officer's eyes as he surveyed me. 'It was either him or you, y'know.'

'Yes, Sir.'

'Ah, you'll get used to it. This isn't a damned picnic, y'know lad.'

'No, Sir. It's not a — er — damned picnic, as you say, Sir.'

Battle incidents like these have their own colourful moments but the other side of the picture is eternally sombre.

Centuries ago, and almost certainly on this very soil at times, heroic Grecian mothers of the Hellenic wars would farewell their sons off to the wars with the customary 'Come back my son — with your shield, or on it', and then go and make secret libations to the gods that it should be with it and not on it.

In the present war, two mothers, thousands of miles apart, equally await the end of the war. In my own little 'two up, two down' dwelling in a Lancashire cotton town, an English mother would be prayerfully scanning the daily and increasingly large casualty lists. Maybe she nips furtively into a church now and again to say a quick one 'just for luck', and awaiting the return of her son.

In some little village in the Anatolian hills, in a tiny one-roomed shack, where the smoke goes out through a little hole in the roof, and where outside are festooned the strings of drying tobacco leaves, waits another mother. But I guess this one will wait in vain. So I'm glad it wasn't my finger on the trigger that would make her wait.

And where's Cpl I—— all this time? We parted company half-way across No-Man's-Land, just after in fact a spew of shrapnel balls from a low bursting shell had torn up the ground in front of our feet. 'I'm going to 'elp that silly young sod over there,' he said, indicating one of our young boys who had temporarily lost his nerve and was stood still, paralysed with fright and screaming his head off.

'Yer on yer own now, Charlie boy. Don't let these bloody wogs get yer down. Just keep a-goin', steady like, and y'll be all right.' His teeth in their temporary haven of his tunic pocket, he bared his toothless gums in a wicked grin, and his eyes glinted with the renewed pleasure of battle.

'Just like old times this, Charlie boy.'

Damn this bloody corporal. He seemed to revel in this orgy of battle, murder and sudden death. With a last call out to me, 'Nil carborundum, tha' knows, mate,' he disappeared among the smoke and dust towards the young lad in distress and that's the last I saw of him. Along with Sergt R——, he got knocked off himself that same night, helping some other poor sod.

And me? I'd managed to keep my nerve, but don't get any fancy ideas about my own particular brand of valour. It's not every young soldier who has the luck to take what could have been his last walk, and almost hand-in-hand with one with such a supreme contempt for the hazards of battle. Part of his massive and unreasonable imperturbability must have brushed off on to me — part of the pollen of his flowered contempt for danger must have sprinkled on me, for I'm damned sure it's no part of my own make-up. All I know is I felt bloody lonely and afraid when he left me.

'Nil carborundum illegitimo' — one of our light-hearted sayings. A Latin purist might reject our free translation, but for us 'it figured.' *Nil carborundum illegitimo* — 'Never let the bastards grind you down.'

And what a grand Latin motto to sew on any regimental flag.

CHAPTER 10

The Harbinger

Reading through all this you could get the impression that I was the only decent bloke in my Company. There's an air of sickening self-righteousness so far that appals me. Let me purge me of my grievous fault.

In the long procession of good officers and NCOs — stretching way back in the history of the British Army — I doubt if the ensuing months ever produced a better bunch than ours. Our Officers, with the same bare rations as ourselves, the same privations, the same bed of the bare trench floor, the same necessity for the thrice-daily delousing act, and yet maintaining that subtle 'above-it-all' bearing which is the hallmark of the good officer. Well-manicured thumb nails crushed between them the same foul little body lice that our own more grubby thumbs crushed. More gentlemanly expressed cuss words consigned to perdition these foul little bastards that plagued us. Our cusses that we used so freely had the coarser and more explosive tang, but the intent was the same. In the quieter moments an officer would occasionally do a spot of 'morale boosting,' sitting chummily with us on the fire-step. In the politeness of good fellowship, we maintained the fiction of believing every word he said and of the optimistic future he painted. The good manners of good fellowship have their own immutable laws.

Lower down the scale, the NCOs did their little best, too.

Their North Country pawky humour lent a welcome mockery to the vicissitudes of life and sustained us in moments of crisis. The art of bullying good humour, so characteristic of the charge-hands in the cotton mills, was now extended in its use to Active Service conditions and was not unwelcome. It had a comforting ring of the sameness of existence — in the mill or in the trenches. Like as if life was just the same, anytime, anywhere. They'd mock at you, these NCOs, laugh at you, laugh with you, even cry with you if necessary, but always with you in real understanding. And if it was like this with the officers and NCOs, it was no less rich with the others of the same humble rank of Private like myself.

Especially with Jimmy N——.

A man of herculean build, solid iron from neck to feet, but solid ivory I'm afraid, quite often, from the neck upwards. A native of the Wigan district himself, miner and son of a miner, he should according to the fashion of those times have enlisted in his own battalion, the 5th Lancs. Fusiliers. He explained his happening to be with us by 'Well, we'd bin watching Wigan playing Rochdale Hornets at rugby. Playing away they was, that's how I came to find meself in yer bloody town of Rochdale. After the match me and my mates was 'avin a bit of a piss-up like, and when I'd sobered up I found I'd joined up in your bloody lousy lot.'

He was perpetually good-natured, always smiling, and with an uncritical acceptance of life's daily botherations that was a source of constant envy to me. Not at all a simpleton, rather a guileless sort of chap, but so one hundred percent genuine that not even the smartest of the smart-alecs in our Company ever tried to take the rise out of him.

Between him and me was a bizarre sort of friendship. He stood in awe of the little scraps of learning I had. Once, during a booze-up in Cairo he heard me carrying on a short conversation at the bar with a French sailor on leave there — me, in my elementary schoolboy French and the sailor himself rattling on at the usual 200 words-to-the-minute speed which is the curse of the Latin races.

'There he was,' he related to the lads afterwards, "im and this 'ere Froggie rattling on, and Cobber talkin' his lingo just like a native.' He forgot to say like a native of Lancashire, for there can be few tortures more exquisite to a Frenchman, even a French sailor, than to hear his own beautiful language murdered in schoolboy syntax and broad Lancashire accent.

But thereafter my reputation with Jimmy as the Learned Oracle was assured. Illiterate himself, he could neither read nor write, but at hands of cards and games of pontoon and brag he could reckon up as quick as any accountant. The onerous duty of reading his letters from home and of replying to them just naturally fell on me. Not that I minded, for this bloke anyway. When he'd see me staggering along the gully on our many trips up and down, and carrying that 80-lb machine gun on my shoulder, he'd take it off me with a gruff, 'Ere, give us that bloody gun, Cobber' and toss it over his own shoulder as if it had been a bag of feathers. He was a strong bloke all right. When I used to ask him what to write home, he'd say vaguely, 'Oh! Just tell 'em we're all right, tha' knows, and all that. Tha knows wot to say better'n me.' So I'd compose a simple letter saying that the food was lousy, but apart from that everything was hunky-dory, and that we'd soon have the Turks on the run. And end up asking for some fags to be sent out. Translating letters from Jim's mum was, however, quite an achievement. I guess their family was more renowned for qualities of heart than for literacy. However, with care, and embroidering a bit from a fertile imagination, I'd read out something that sounded plausible. Jimmy used to sit on the fire-step, expectant-like, and with a happy smile on his face, eagerly awaiting the news from home — much like a kid waiting for the curtain to go up at a concert. Goddam me for a louse-bound bastard, the little lies I'd invent, just to see the happy grin on his face, and even if it was all done with good intent.

His two brothers were serving in France in the same battery of artillery. In the '14 war, if you were brothers, you had the right to claim to serve together in the same unit, so these two brothers enlisted in the same battery soon after the war started. Jimmy was never tired of telling us little anecdotes of these younger brothers of his, how he'd taught 'em to scrap, etc. 'They could beat anyone of their own age for miles around,' he told us proudly.

One grey mizzling day in November later that year came a letter from Jim's mum I had to read to him, and I'd have swapped my hopes in Paradise to have side-stepped the job. Jimmy waited patiently as always for me to digest and then translate his mum's epistle. Growing restive he gave his good-natured guffaw and asked 'did I want a pair of specs or summat.' I lacked the guts to tell him. The job had to be done, but damn the fates that picked me for this particular one.

'It's bad news, Jimmy,' I said, and he tensed up, sudden-like.

'Which one is it?' he asked quickly, 'John or David?'

He knew from whence to expect such news, there's no doubt. So I read on:

> My dear son,
> I am sorry to hav to let u no that David and John hav bin kild.
> They was kild the same day wen a shell hit their battery. My hart is broke.
> Your loving Mum.

Jimmy sat there a while, uncomprehending, then turned on me, his eyes blazing. He held my wrist in a grip that nearly cracked the bone and said pleadingly:

'Tha'art kiddin', Cobber, Arn't tha'? Say tha's kidding. Tha'art 'avin' me on, Arn't tha'?'

Sadly I assured him that I wasn't having him on, but he still found it hard to believe and snatched the letter from my hand and passed it to a Sergeant who stood nearby cleaning his rifle.

"Ere, Sarge, read this, will yer, mate. If Cobber's 'avin' me on I'll break every bone in 'is body, that I will.'

The Sergeant demurred and passed it back to him.

'Nay, Jimmy, if Cobber's read it for thee, I guess it's true.'

I myself passed the letter over to the Sergeant. He was some time deciphering the unique style but at last he mastered it, gave it back to me, then turned his back on us, fiddling noisily with the bolt of his rifle.

Jimmy sat there for a time, frozen to stone and I left him quietly and told the other lads. Their re-action was normal. Someone had just brewed a mess tin of hot tea, and together we drained the remnants of the rum we had stored among us, lacing the tea until it was more rum than tea. One of the lads sidled up to Jim and laid it alongside him with a gruff, "ere, drink this Jimmy lad, it'll do thee good.'

CHAPTER 10

To a mate in trouble the solicitation of rough soldiers is about as nice a thing as you could wish to see, this side of the River Jordan anyway. But the tea stood untasted and grew cold.

'Summat's gone wrong wi' Jimmy,' one of our chaps said some days later. He certainly was a changed man. 'Summat' had gone wrong, too, with our erstwhile friendship. Rarely now he'd exchange a word with me. He'd still snatch that adjectival gun from off my shoulder when I was treading my own private 'Via Dolorosa' as we moved about from place to place, but he always took it without a word.

If there's a moral in this story, and there doesn't have to be, but if there is one it is that to be the harbinger of bad news is to be on a bet to nothing.

CHAPTER 11

A University Degree

This bit of trench we had won was won with surprisingly light casualties. A close boyhood friend of mine, Frank Whitfield,[8] was next to me as we went over the top and a few yards out was shot in the temple. Son of a Rochdale police sergeant he was, like me, very keen on boxing as a lad, and twice a week in the Police gym. There we'd belt hell out of one another. We were about evenly matched, and only the outbreak of war stopped us ever finding out who was the better man. When I saw him bowled over I thought the question had been decided for good. Close as we were in friendship, I couldn't stop to help him — at such times you can't even help your own brother. And we'd never know who was the better with the gloves. Or so I thought.

On leave myself in my home town just after the war, I was astonished to bump into him, alive and well, and with three pips on his shoulder and the RFC pilot's wings. Miraculously he'd survived his early wound and with no ill effects. The very last I saw of him was later that same night after we'd spent the evening in the 'Golden Fleece', discussing old times and comparing subsequent events. I am a bit hazy as to what happened later in the evening, but I remember the scandalised landlord of the pub, assisted by the barman and two friends, hustling us into a waiting cab about half an hour after closing time. Well, it wasn't our fault really; it was the landlord's own fault for selling such strong beer. The cab dropped us off at Frank's place where I spent the night, but

---

8 Probably Sgt 8613 Francis Whitfield. See Appendix II, Notes on individuals mentioned in the text.

when I awoke the next day he had gone, catching an early train back to his unit where he was already overdue from leave. When I called at The Golden Fleece the next night I was definitely persona non grata with the landlord for a few minutes, which puzzled me. But apparently after closing time the night before, Frank and I had wanted to settle the question as to who was better with the gloves, and in the Saloon Bar of all places. But after the landlord recognised me as the son of a man who must have been one of his best customers, we got on famously.

Although we'd won that bit of trench fairly easily, we had to struggle hard the next two weeks to hold it, for Johnny Turk is a doughty fighter. Many times the rifle barrel and stock got so hot to the touch that we bad to hold it with a bit of thick cloth to avoid burning our hands, and the ground in front of us became littered with Turkish dead. By the end of these two weeks we were about all-in. It had been an endless task in the daytime and at night a continuous shift of sentry duty on the fire-step — one hour on, one hour off, in turn, right through the night. Ask any soldier who's done a couple of weeks like that — ask him what it's like — but before he answers and if you're a bit squeamish on obscenity, first stuff your ears with cotton wool. So, when we stumbled along the gully one night a fortnight later on our way to the beach on being relieved by fresh troops, we were exhausted. In the picturesque littoral of the cotton towns, we were 'bloody well on our knees.' On our way down we passed the troops relieving us, New Zealanders just landed that day, full of starch, self-confidence, brash, bronzed and healthy — not like us, wan and forlorn. One of our chaps called out as we passed, 'Give 'em hell, lads, show 'em what you can do.' They called back, cocksure and confident, 'Just wait till we get at 'em; nobody knows what we'll give 'em.'

We relieved these New Zealanders some ten days later.

Chastened and quiet, they passed us glumly without a word on their way back to the beach for a rest.[9]

On the beach we had a marvellous rest there, in the lee of the cliffs — most of the time spent in catching up on lost sleep. Even the rations were a little better but we craved for tobacco. On the lighter side of life there were amusing incidents on the beach even though at times

---

9 See Appendix III, Notes on events and places mentioned in the text.

it is prudent and kinder, before talking about them, to creep under the blessed shawl of anonymity. You never know what relatives might still be left alive of these chaps. So we'll postulate a mythical company of a Lancashire regiment enjoying their well-earned rest. One of this gang had managed to 'nick' a carton of 5,000 cigarettes destined for the 'higher ups' whilst he was on a stores fatigue party. The 'Nicker-in-Chief' was a colourful character. In civil life he was a scrap dealer or, as what we used to call them in those days, a rag-and-bone picker. But what the hell — a rose by any other name smells the same, I guess. And what an old-time rag-and-bone picker didn't know about the gentle art of misappropriation could be written on the back of a penny stamp. A rough diamond all right, this one; they didn't come any rougher, but a staunch pal to his mates. Resourceful, cunning and unscrupulous, but to his pals faithful unto death. Graduate of the University of Life, he held an Honours Degree in the Science of Illegal Misappropriation, and a Research Fellowship in Industrial Psychological techniques and their application to Life's tight corners. Expert pickpocket too, his touch was light and deft. While you yawned he could abstract the gold filling from your teeth, and you'd feel it no more than a butterfly's kiss. These fags he evenly distributed among the surviving members of the Company, with about 50 fags kept in reserve, for — as he said — 'you never know wot 'appens.' In this tobacco-less Eden (!) these fags were as precious as life itself. No wonder the high-rankers for whom they were destined got a little peeved. There was a hell of a bloody row — couldn't have been more fuss if someone had pinched the Crown Jewels. An order went out that all kits were to be searched, but thanks to a timely grape-vine warning the fags were 'cached' in a little hole in the side of the cliff, with just a few kept hidden handy for a crafty smoke until the row blew over. It was during the savouring of these odd fags the same night that Nemesis nearly overtook the delinquents. From behind the ground-sheets stretched tent-wise over the little one-man dug-outs cut in the side of the cliff, the scent of cigarette smoke rose maddeningly on the air, attracting the attention of an officer doing his rounds. Too late to destroy the evidence but not too late for our resourceful 'Nicker-in-Chief.' He approached the officer smartly, chucked him up a posh salute and produced a packet of 20 with the tell-tale identification marks on the packet.

'Beggin' yer pardon, Sir, I was just comin' to see you. I found these 'ere cigarettes near the incinerator today when I was burning a bit of old rubbish, so I thought as 'ow I'd better 'and 'em in like. The lads wanted me to share 'em out, but I says, No! That wouldn't be right. There was one or two loose ones lying around wot I gives to the lads, but this 'ere packet, like — I says — best give 'em to an officer, I says. He'll know what to do with 'em.'

His look of righteous oiliness as he handed over the packet was worthy of Pecksniff himself.

'You did quite right, Private... You did the correct thing. I'll hand these in to the Quartermaster.'

'That's just wot I says to the lads, Sir, says I. An officer will know best where to 'and 'em in, or wot to do with 'em.'

No doubt this cigarette-starved young officer would have known what to do with them. From the rich store of his Applied Psychology, the Nickerin-Chief dropped another gem — 'Once you've got 'em with you, you're safe' — whatever that cryptic remark might have meant.

Three months later the Nicker-in-Chief got 'his' from a shrapnel ball, over-assiduous in his hunt for Australian blow-belts and Australian gold. Not the first one, this chap, in the history of our race to come to grief 'driven clean mad for the muck called gold.'

His loss was a great blow to us. We'd never see his like again. Some of the lads thought he should be canonised!

I wouldn't altogether go along with that myself, but when you come to think of it, and in these more modem times, his unique gift of appropriating what doesn't belong to him would have made him a wonderfully successful train robber... or an outstanding Chancellor of the Exchequer!

CHAPTER 12

The Numbered Bullet

What makes for personal survival in war? Luck? Fate? Or the sixth sense telling you the split second when to duck? Or a combination of all the factors?

I still don't know. A rifle cracks a few hundred yards ahead, and you're grateful the speed of sound exceeds the speed of the bullet — even if only just. And the bullet merely parts your hair instead of messing up your brains. You've ducked just in time and lived to survive, maybe, a hundred other such hazards in the future. I said 'maybe'!

The Goddess of Luck sits in for a hand sometimes and the game is played again. The ricocheting bullet by sheerest chance hits the metal cigarette case in your left breast tunic pocket, and all you get is a minor thump over the heart and for once in your life you're glad you're a fag smoker, in spite of the morning cough from those damned insidious white coffin nails. And you've got now a messed-up cigarette case, battered and bent, that'll be worth any amount of free drinks in the pub back home after the war, and when you start telling your old soldier's tales. So maybe it's a fair swap.

The ominous Black Ace of Spades comes up sometimes from the pack and you know that Fate sits in for a hand, and you can't do much about it.

As the wise old man says to the boy Buddha, on his first glance of the incidence of Fate and Death:

> A stumble on the path, a taint in the tank,
> A snake's nip, half a span of angry steel,
> A chill, a fishbone, or a falling tile,
> And life is over, and the man is dead.

'For who can shut out Fate.'

'If yer name and number's on the bullet, it'll find yer, mate, and yer can't do much about it,' says the soldier. Me — I dunno. All I know is that some chaps do have the damndest luck. Like Cpl Grimes of our gang.[10] (The name's fictitious — and if you like let's say the whole incident is fictitious — you'll have to judge for yourselves.)

'For who can shut out Fate?' Or was it just sheer bloody stupidity, rank carelessness, or just the exuberance of youthful high spirits on our part? I guess we were all a bit to blame even if we hate to admit it. Maybe if you're a strict moralist you might think it was retribution overtaking one who had gleefully broken the seventh of the Ten Precepts permanently inscribed on stone for our eternal guidance. But in wartime, and especially in the Middle East where sex is rampant in the very air you breathe, adultery doesn't seem the same. It doesn't exactly break the Seventh Commandment — merely bends it a bit.

'Thou shalt not kill.' We seemed to be breaking the Sixth Commandment with impunity, even with strong official blessing. So if our moralities got a bit mixed up in those war years, maybe the fault wasn't entirely ours. Anyway, if you're a bit strict on moralities, maybe you'd better skip the rest of this chapter.

After our few days' rest on the beach, we had another fortnight back in the same trench. It turned out to be a quiet fortnight. It gave us time to adjust ourselves to trench life. We even came to like it — this primitive call to live deep in the good earth, like as if we felt an age-old pull back to the shelter of the cave, back to the shelter of the hole in the ground. We even started getting finicky about the tidiness of our own individual patches. Blokes would dust the loose earth off their patches of the firing step with an empty sandbag, like fussy old women keeping the house neat and clean. Out of sheer boredom we burnished our mess-tins till they shone like silver. Officers started rifle inspections and we got torn

---

10 See Appendix II, Notes on individuals mentioned in the text.

off a strip if there was a speck of dust in the rifling of the barrel — all this in a life semi-underground. 'Just like bloody peacetime again,' one of our blokes grumbled, after an officer had spotted a microscopical speck of dust in his rifle. The odd shell would burst over us occasionally, smothering our nice clean patches with muck and debris, and all we felt was the grievance of the housewife when somebody's trod all over her carpet before wiping his feet. But nary a casualty during the whole fortnight — both sides seemed temporarily to have shot their bolt.

Not quite like peacetime, but it gave us leisure to discuss the really important things of life — like the merits and demerits of our own local rugby team. Anecdotes of courting episodes of the cotton town — anecdotes that would have brought a modest blush even to the fair cheeks of Lady Chatterley — were swapped around, for we were at the lustful hour of Life's morning. Married men of our gang relived volubly old marital delights and solaced themselves with threats of further delights to come 'after the war.' Lacking the finesse of innuendo, language was colourful and often brutally frank. Photos of wives and kids were passed around sentimentally a hundred times. The trivia of life, such things as death, mutilation and destruction were properly relegated to the insignificant.

Probably the most surprising and hopeful characteristic of mankind is his fantastic capacity for adjustability. The life we were living seemed to be the life we had always lived, the only life possible for sane men to enjoy.

At night it was right and proper that we should spend the dark in man's proper night routine on sentry watch, leaning with the rifle over the parapet — now and again ducking for a few seconds until the Verey light had shed the last of its baleful luminosity, or occasionally letting off a few shots enemy-wise, at random. As a lone cur at night, startled by some nocturnal shadow will start a chain of answering barks all around the town until the whole chorus dies of inertia, so a few shots fired at random would bring the whole line into a continuous fusillade of vicious firing, eventually to die away when tired of it. And that's the pattern of trench life at night in the first war, whatever the theatre of battle.

'That's right, you bastards, keep quiet for a bit. I was just telling you mate, wasn't I, about that bint I had in Alex one night... oh! There they go again, the bastards... trigger 'appy, some people, mate, that's what they are, trigger 'appy. Well, like I was telling you...'

But the two hours sentry stretch comes to an end and you bed down luxuriously on the bare earth of the firing step, noise and yarns forgotten in dreamless sleep. The hard earth is as soft as a feather when you're tired, and the continuous fresh air is a chloroform pad. At dawn someone gives you a friendly kick to pull you awake for the dawn 'Stand-to,' that mythical period of exceptional danger that is legendary noted for being fraught with the risk of surprise enemy attack. Whereas, in truth, both sides are propping up their tired eyelids with matchsticks. Old army traditions die hard. I guess this particular one dates from the days of armies in the open field of Wellington's day. If you like, it's a bit like the present East and West arms build-up, both afraid of the other and both reluctant to start anything, and especially before breakfast.

Hot tea, minus milk and sugar, after 'Stand-to,' and the circulation coursing in your veins again. Loud greetings of friends, the morning scutter to the latrine hole as the hot tea begins its beneficial intestinal scratching, the raw hunger for the non-existent fag. And another day begins. In the warm Mediterranean sun, shirts come off for the periodic delousing act and for new exercises thereby in profanity. Still, this is man's normal life. It seems like we've been doing it for thousands and thousands of years. Almost we are unable to visualise any other sort of life pattern than the present one. We all agree we hate every minute of it, unconscious liars that we are. At the back of every man's eye is a new gleam of awareness. Maybe the gleam is only discernible to God Himself. It's the new joy of living. Our lives are now spun out on the tightrope of Life and Death, and boy, oh boy, isn't every precious step we take on that tightrope damned exhilarating. Without realising it was had discovered the joy of living on borrowed time, of cherishing every moment of its preciousness. Just the joy of possessing a little bit of Life itself we'd found, and what we had found we found to be good. Every hour a new meaning, every moment a new exultation. Yesterday had gone; tomorrow was most unlikely — for us at any rate. Tomorrow was yesterday's today. Today was yesterday's tomorrow. No tomorrow's cares to take care of, only today, the precious *today* of the primitives, the fleeting, alive, present moment we cherished so gratefully.

Ay! But what would I give just to live once again in an unheeding Timeless Universe.

The quiet fortnight ended in this trench; we were told to get ready to move that night. It was not to be to the beach for another spell of rest but to occupy another part of the front line abutting off the Western coast line of the Peninsula, a part of the battlefield at present unknown to us.[11] Just before midnight we filed out and oddly enough were assembled in groups in the open. We were to be guided to our destination by a young Greek boy familiar with the local topography, a sort of official guide, dressed only in a badge-less green suit and a bright smile. He seemed to be simply revelling in the job, laughing from ear to ear, his white teeth flashing in the dark as if he roared at some hidden joke. Darkness bothers me not at all, in fact I find it strangely comforting, but there were amongst us many normal and brave chaps in the daytime, but who at night were as terrified as kittens. This fear of the dark, it's something you've got or you haven't. In any case, it's something we can't help if we have it. These blokes grumbled a lot at the dark and at the danger of cross-country moving in the open, but our guide knew his job. At times he'd signal us to keep dead still for sometimes as much as ten minutes at a time, then cautiously we'd creep along again. It reminded me very much of some two or three years back in the Scouts, back home, and playing just such similar games. In fact, but for the stinking smell of death that always lingered on this land, you'd swear it was nothing but another Boy Scouts game.

But I guess we had to do it this way. There was no sheltering gully leading to our new destination. Fortunately there was a strong down-wind from the Turkish lines. Damn good job, too, that there was a strong down-wind for you'd have sworn that the clatter from our mess tins and other accoutrements would have wakened the dead. It took us about 3 or 4 hours of apprehensive travel, yard by yard, but eventually we got there about an hour before dawn and literally fell asleep in our new trenches. The usual orders were given for NCOs, and sentries to be organised into night watches, but that's about as far as it ever got. A shame-faced duty officer rolled up about 10 a.m. instead of

---

11 The Fusilier Bluff firing-line. See Appendix III, Notes on events and places mentioned in the text, and map 8.

at dawn to awaken the equally apologetic Sergeant of the Watch, and a whole shower of sleeping sentries. By god! But we must have been tired. By unspoken gentlemanly agreement the whole thing was tacitly ignored, except for some disgruntled 'old sweat' who had been sleeping as soundly as any of us. 'We could 'a bin murdered in our beds we could an' all.' His mate agreed, adding drily, 'That is, if we'd 'ad any beds.' Good humour restored, we appraised our new home appreciatively.[12]

'A soldier's Paradise' one of our blokes christened it enthusiastically. Well, it was OK by me. The trench was set on a hillside running steeply down to the sea below, sparkling in the morning sun. The configuration of the land in front — the No-Man's-Land — was such that an attack was extremely unlikely except in a major engagement. Myself, I could have stayed there for a thousand years, with that sun, that glittering sea below, and the comforting view of the British Navy ringed out at sea at anchor, some two miles out, this ever-present view of the good old British Navy, this brooding watchful Mother, always in comforting sight of her feckless children on shore. After we'd mildly scrapped amongst ourselves for favoured positions in the trench, dumped our packs and belongings, we were as happy as kids at a holiday camp. Even the grub was more plentiful — six biscuits a day instead of the usual three, and a whole tin of bully-beef, regular, every day. If only we'd had some fags we wouldn't have swapped it for a king's throne. Still, you can't have everything in life. In batches we'd go swimming in the Med., savouring the warm enveloping of its magic water on our grateful skin. It was all right as long as you didn't go too far out, for the configuration of the land protected you from machine gun and rifle fire, but if you ventured too far out the sea around you would be peppered with bullets till it looked like someone had scattered thousands of pennies in the water. Once caught out too far from land, I managed to get back OK, but half dead with fright. Scant sympathy I got from the Sergeant in charge. Heavily sarcastic, he greeted me.

'Me and the lads was just thinkin' of collecting for a wreath for thee, tha' silly young bugger.'

Then, very angry, 'If ever Ah catches thee doin' that again, Ah'll put thee on fatigue all day, that Ah will.'

---

12 Fusilier Bluff.

He needn't have worried. If ever I did that again, I guess he wouldn't have to bother, for you can't do fool things like that twice and live to tell the tale.

When we weren't swimming, we'd spend the time arguing or discussing what we'd do after the war. All agreed we could never go back to indoor work again — 'not after this lot.' Mercifully denied the gift of second sight we were unable to see ourselves years later, scrabbling for any old job, indoors or out, in long endless queues, and getting hardened to the snooty superiority and sneers of the wise boys who'd dodged military service and collared the best jobs.

Relaxed hours, the war forgot, and the warm sun — the novelty of the present life — it all tended to induce verbal intimacies not possible under any other conditions. Sentimental feelings just naturally ran riot in the enforced idleness.

One of our chaps, Cpl Grimes, a married man with two kids, always kept a woman's worn cotton stocking folded on top of his belongings. At times we'd observe him finger it longingly but we tried hard not to see this damned stocking. The rougher soldiers are the more delicate they are not to pry into a man's sentimental privacy, especially domestic privacy. Even scallywags like us, to whom nothing and nobody was sacred, would respect a man's domestic privacy and the reputation of a man's wife. If we thought about that damned stocking at all, and we tried hard not to — it might have been a tenuous link with his wife, a memento of other happier days. Nothing unusual in that — to us. We Lancastrians allow sentiment to have an unusual place in life and are openly unashamed of it.

During the usual sultry morning natter one day about this and that, a gust of wind dislodged this damned stocking, depositing it near the feet of one of the young lads nearby. With an air of awkward unconcern, he picked it up and gave it to its owner. Cpl Grimes thanked him gravely then burst out laughing.

'Did Ah ever tell thee, lads, 'ow Ah got this bloody stocking? One of the bints in Cairo give it to me — well, not exactly give it to me, but sort of — if you know what Ah mean.'

This was a different kettle of fish and we crowded round to listen, grateful for the break in life's monotony. A sea of boyish eyes glowed lustfully in anticipation. Cpl Grimes turned to me, "Ere, Cobber, tha's too young for this — away and get thee mi tobacco tin.' I was off and back in a flash and caught the tail end of the story.

'Well, like Ah was telling you, we'd bin on detached duty 10 kilometres north of Zag-a-zig. Telephone duty it was, one hour on, one hour off, in turn — just me and my mate during the day, but a good night's sleep every night. Two months on end we'd bin on that duty, all on our own, me and my mate, and wi' the ration wagon calling on us twice a week. Nowt to complain of really — all the rum and fags we wanted and the grub was good too. But a bloody lonely job it wur, nowt to read much, no-one to talk to, nowt to talk about except the bints we'd 'ad in Cairo and Alex, and wot we'd do to 'em when we got back. Ee, but 'e wur a bugger for 'is women, wur my mate. But it seemed like we'd never leave the bloody place, till one day the relief team turned up and we 'ops into the train at Zag-a-Zig about 6 a.m. and gets to Cairo about 8 o'clock. I don't mind telling yer, lads, wot Ah was thinkin' about on that train. I could see nowt before me on that bloody train but a bint in front o' me, and wi' nowt on — well, yer knows wot it's like, lads, when you've bin on the desert for a couple of months, and wi' the old sand under the foreskin.'

Amid the laugh one piped up, 'Nay, but we don't, Corporal. Tell us.' — And laughed his head off. But Cpl Grimes told him rudely to shut his gob while he finished the yarn.

'Well, we gets to Cairo like Ah said, me an' my mate, and knocks on the door of one o' them houses I'd bin to before. A big fat Madam comes to the door and tries to shoo us away like. "Come back 2 o'clock," she says, "not open till two."

'Oo, 'orrible she was, big fat bitch with 'er 'air in curling pins and looking like she'd like to strike us dead. But me an' my mate wasn't 'aving any, and we pushed 'er to one side and goes upstairs. We couldn't see nowt nor nobody — we reckoned all the bints was asleep or summat. But then we hears a noise like water splashing somewhere. So we goes round the corner and I opens the door of a little cubby hole where two bints were taking a shower.'

Cpl Grimes paused, lost in rapturous reverie. 'The one I picked, little Syrian girl she was, no higher than two pennorth o' copper — about five foot, if that. Little dark brown girl she was, wi' black fuzzy hair all over her head — like these Syrian bints 'ave. Lovely figure she'd got, too;

and her flesh, well — she looked like she was made of soft India rubber. Bloody luvly she was.' Cpl. Grimes gave vent to another long ecstatic sigh, then resumed.

'Her 'air wasn't wet and sloppy like you'd expect after a shower, that sort of 'air doesn't seem to hold the water. Covered in millions o' bright little diamonds, her hair seemed like — little bright diamonds from the water drops, if you follow me. Like a chocolate coloured fairy princess crowned in diamonds, she looked to me. She did, lads — real beautiful she looked. 'Ee, but she did look grand. I nearly forgot wot I come for, she looked that luvly. When she sees me standing there she lets out a squeal and gives me a saucy grin, and I picks 'er up, all wet and cool she was, and me all burnt up, like.'

Cpl Grimes paused to relight his pipe, and looked round slyly at the sea of lustful young faces.

'Yes, go on, go on. And then?'

But Cpl Grimes gives another aggravating grin and says, 'And after that, lads, me and my mate goes to one o' them little Gippy Caffs and 'ad a bite o' breakfast and a wash and a shave.'

Blessings on ye, ye naughty wicked little girls of Cairo and Alexandria. Denizens of the Wazar Berka, Cairo, and the infamous Rue des Soeurs, Alex., you plied your wicked, joyous trade. Some day in a more enlightened age they'll erect a monument to you. Target-less sex, in an army of young soldiers at the peak of their sexual maturity, leads only to the canker of homosexuality. You prevented us, maybe, from drifting into the pattern of what legend whispers of public schools and the darkened alley walks of ships of the Royal Navy. Because of you, you wicked little fairies, we never grew up to become what is known in modern parlance as 'consenting adults.' Your very sensuality, your brazen deplorable primitiveness was in a way as innocent as child savages. You'd flaunt your wares on the door steps of your houses, and often on the main street, openly, clinch a sale with the hesitant customer, deliver the goods speedily and efficiently, and snap the proceeds of sale — a mere 5 or 10 piastre note — safely into the oldest bank in the world, the stocking top.

Or so the boys told me!

Goddesses of the 'short-time' — was ever such phoney joy and momentary relief obtained so cheaply. By your very existence and by plying your reprehensible trade, you also made life safer for your more virtuous sisters who scorned you so bitterly. I doubt during the whole time of the occupation of Cairo by British and Colonial troops if there was ever one case of a respectable daughter of the local inhabitants being raped, assaulted, or even accosted — thanks to you, ye wicked little gals. The very people who scorned you — the local inhabitants who were powerful enough to have swept you out of existence — they tolerated you, and were so often your best customers. They tolerated you — not, I fear, in Christian love and charity, but rather to make life safer for their own wives and daughters.

The yarn finished, Cpl Grimes stood up and knocked out his pipe, and reached for the tawdry memento of love.

"Ere, gimme that bloody stocking. 'Appen Ah'll give it back to her one day.'

He stood pensive for a bit, a tender smile playing oddly on his commonplace features — like as if remembering less lurid moments of this thing called Love. But the ape-like activities of sex oft leave in their wake more fragrant memories. This thing called Love — is it from the gods, or whence?

Re-action now sets in and boredom grows. In sheer desperation and anti-climax, someone started 'playing silly buggers' as we used to call it, doing the daftest things, things like snatching off a cap from someone's head and tossing it over the back of the trench. This started a free-for-all cap-snatching till we got tired of clambering over the back of the trench to retrieve our headgear. About ten minutes later someone suddenly asked, 'Where's Cpl Grimes?' Nobody knew.

'Last time I see'd 'im,' said one of the lads, "e wur getting 'is cap back, out yonder.'

We jumped over the back of the trench. Well, you never know, he might have sprained his ankle or something and couldn't get back. Rough country like that could rick your ankle before you could say 'knife.'

We carried him back. Sorrowfully, regretfully, we laid him down and cursed the sniper or the stray bullet that had ended his short life on earth. But maybe by now he's found a Mahomeddan paradise of black-eyed houries, and 'little Syrian bints wi' their hair crowned in diamonds from the little drops o' water.' I hope so. He was a decent bloke to his mates.

We counted him lucky that his end must have been quick and painless — shot clean through the heart.

Near where his kit lay neatly stacked, we hollowed the trench floor a couple of feet and stamped him firmly down. The burial service was carried out by the Sergeant, who said the Lord's Prayer over him — at least as much as he could remember of it, and we all joined in a ragged 'Amen.' In those early days there was no official burying place — a man was either left for his flesh to blacken in the sun and his bones to bleach, or hurriedly pushed into a shallow grave.

Before we started shovelling the earth over him, one of the young lads reached out for the stocking and dropped it reverently on the chest of the deceased. Depressed and hushed, we sat around for a bit, resurrecting anecdotes of the unfortunate Corporal's hitherto unnoticed virtues.

The Sergeant nodded approvingly at the lad who had dropped the stocking in the grave. 'That wur a nice thought, that wur, lad,' — and there was a chorus of approval all round from us.

Later, 'Wot made yer think of it, lad?' asked the Sergeant, curiously.

'Think of what, Sarge?'

'Y'know — that stocking.'

The lad wriggled uncomfortably a bit. His lower lip trembled and he lowered his head. In a choking voice he blurted out, 'Well, like 'e said, 'appen he'd want to give it back to 'er one day.' Then he burst into tears and rushed away.

We sat around for a bit in silence — the high embarrassment of men caught up as unwilling spectators in a tragic emotional outburst. Our eyes sought the Sergeant hungrily — the older man who never failed to produce the right slant on things. We badly needed a conductor to discharge this high electric field of emotion. But for once the Sergeant failed us. Lost in thought he was squatting on his haunches, staring at the little mounds of earth his desultory fingers were fashioning on the trench floor. All he could utter was, 'Poor little sod.'

Then he stood up straight, half turned his head and shouted angrily to the empty air, 'They've no bloody right to send young kids like that to war.'

CHAPTER 13

Compensations of Eternity

It's very odd how a lone casualty in wartime should stick out like a sore thumb. You can lose a hundred or so of your mates in the heat of an engagement and the next day the whole thing gets blotted out of your mind by the very immensity of it. The very size of the casualties seems to produce the mind's own anaesthetic. I wonder if it'll be the same when the Big Bombs drop in the fateful future!

But a lone casualty during a period of comparative calm is, I guess, a bit like the unexpected death of a comrade in peace-time. And maybe the thing was accentuated by the unaccustomed chore of having to carry the rifle and kit of the deceased Corporal back to the Company Quartermaster's store. And that's a thing we never had to do, ever, for any of the other blokes knocked off in battle. Their rifles were left to rust in the dew-wet grass, and the mice and rats found new and exciting homes in the kit and accoutrements of the chaps lying about in the open — sharing, in the enforced comradeship of war, their homes with the unprotesting blokes already in possession.

Another chap and myself were detailed by the Sergeant to hump poor Cpl Grimes' belongings back to the Company QMS Dump, some half-mile back. But what the CQMS would do with these things is anybody's guess. In carting these things there at all, maybe we were only unconsciously following a routine peacetime pattern. It's a psychological puzzle, but with no marks for the right answer.

A snug little hidey-hole had the CQMS — tucked away safely in a protecting fold in the terrain. Large tins of Army biscuits — square shiny tins of about two-foot cubic measurement and neatly laid like bricks for the walls, made a cosy little house for the Quarter-Bloke, and with a tarpaulin stretched over the top he was secure in all weathers. The walls of this little tin-box brick house were hot to the touch from the strong, sun, but inside it was beautifully cool and dark.

'Shouldn't mind being a Quarter-Bloke myself,' our Sergeant said longingly.

The Quarter-Bloke opened a jar of rum, and we had a tot, toasting sadly the absent Corporal. The Quarter-Bloke was really worried what to do with the kit we had brought in.

'S'pose you'd better leave it here,' he said in grumbling acquiescence.

The personal effects — the cheap little wallet containing the photos of Grimes' wife and kids, and the last letter from his wife and the Identity Discs, all these the Quarter-Bloke handled reverently.

'I'll give these to his Platoon Officer tomorrow; 'appen 'e'll want to write and tell 'is missus.'

And the Platoon Officer will have once again another melancholy job of consoling Mrs Grimes in her new status of widowhood, and of her fatherless children. Tactfully and softly would he lessen the blow with the conventional, 'You will be proud to remember always that your husband died doing his duty nobly for his King and his Country.' Dying for King and Country includes a whole host of things; it even included 'playing silly buggers,' like we were doing at that unlucky yesterday. And Mrs Grimes will look at the colour-tinted photograph of her uniformed husband on the mantelpiece with a new bewilderment. The 'touch-up' art of the photographer and the sad glow of pride she feels after reading the Platoon Officer's letter — they all lend a hitherto unnoticed nobility to his commonplace features, but she's still bewildered that he should have left her alone in this world, like this, and with two nippers to bring up on her own — and what with the high prices that things are in the shops these days in wartime.

At the Most Supreme High Court, Cpl Grimes' Defending Counsel will himself be a Lancashire-born man, for you get the right to choose your own counsel. And although the Counsellor himself had been 'up there' for many years now, he was still unable to shed his Lancashire accent and his Lancashire dialect. They say you never can.

CHAPTER 13

'With your Lordship's permission, these soldier lads, like, they're nowt but a lot o' kids, reely. Kids playin' at bein' grown men, like. Not that Ah'm defending mi client carrying on with that Gippy bint. Him being married an' all — 'e didn't ought to 'ave gone laikin' about wi' them Gippy lassies. But y'know what it's like in Cairo, m'Lud — hot as... well... hot as t'other place, y'Lordship,' etc., etc.

But the Most High Chief Judge was stern, and wasn't falling for this line of defence.

'Eloquent as your plea may be, Mr Counsellor, it still doesn't excuse Cpl Grimes. In any case, what had this Egyptian girl got that his own wife in England hadn't got?'

'Well, y'Lordship — now you cum to mention it, this Gippy bint hadn't got nowt that his own wife didn't 'ave. Ah s'pose the only difference was that this Gippy bint had hers with 'er, nice and handy, like.'

While down below, on this little shell-torn, war-wracked Peninsula, bathed by the calm Mediterranean Sea sparkling in the sunlight, the Quarter-Bloke was still fretting about Cpl Grimes' kit.

'What to do with all this bloody stuff, Ah'm buggered if Ah know.'

Plainly the Quarter-Bloke preferred blokes who if they had to be so unfortunate as to get killed, they should get killed in action, in the open, and without all this returned kit nonsense.

But there's no doubt about it — the Quarter-Bloke was genuinely shocked. He'd been fairly close to Cpl Grimes when they were both corporals together in Cairo, and before he himself became CQMS. He drained his tot of rum sadly, and replenished.

'Poor old Grimes,' he said piously. 'Cor! Many's the night out we've 'ad together wi' the bints in Cairo.'

In a hushed voice he resumed, 'Well, 'e's had *his* last bit o' crumpet.'

'That's more than anyone can say,' said the Sergeant stoutly. 'He's always bin a good soldier, and a good pal to his mates. Who's to say 'e'll never get another bint?' After a pause, 'h s'pose there'll be some lassies in Heaven, or wherever it is we goes to?'

The Quarter-Bloke pondered on this for a bit. 'Ay, ah s'pose there will, but,' regretfully, 'not Gippy bints. 'Cos they're heathens like. Ah reckon that sort doesn't get to Heaven.'

But the Sergeant still defended the prospect of future bliss.

'Oh! I dunno. Parson says we all go theer sometime — not only English but everybody.'

'What! Everybody — as well as English? Wops and wogs, and such like?' The Quarter-Bloke was plainly unbelieving.

'I don't see why not,' said the Sergeant, whereupon the Quarter-Bloke appeared to brighten considerably. He licked his lips appreciatively.

'Grand lassies, these Gippy bints,' he said. 'I knew one once as could make it last nearly as long as a football match.'

* * *

LOST ENDEAVOUR

PART II

(A hotch-potch of Gallipoli memories)

CHAPTER 14

The Cross and the Crescent

The first few weeks of painful initiation over, we had settled down to a routine that varied little day by day. Mentally we had come to terms with our new way of life, patiently, stoically and hopefully. The war we had once planned to finish by Christmas 1914, we now put back to Christmas 1915. Or say, in about six months' time. For sure, by then, we'd have driven the Turks off the Peninsula and be home in England in time for Christmas.

I even came to love this two-bit piece of land that Fate had decreed should be so sanguinary a battlefield. I'd waken in the morning, greet the warm sun with pleasure, hug a secret joy in being a participant in this technicolour saga of adventure, stifle the little misgiving that I might not see the sun set the same day, and cast my eye with a proprietary air over what I now considered to be my own private domain, a domain reaching from the heights of Achi Baba, right back to the jagged beaches, off which lay the comforting line of the light grey ships of the Mediterranean Fleet. Watching those beautiful ships gave me the same pleasure as a kid gloating over a box of new shiny toy soldiers. The flagship held the stage — its big guns visible and menacing, and in respectful attendance were the light cruisers and destroyers, all of them motionless and still on that calm sea. As the poem says, 'As idle as a painted ship, upon a painted ocean.'

It was like watching a framed canvas in oils, these dead still ships on the calm waters of the Mediterranean lake, and the frame, the frame was the background of the blue sky. The only signs of life on the water were the ships' heliographs catching the sunlight in their mirrors as they winked out their mysterious dot-dash messages to one another. The only signs of movement on the sea were the little minesweepers, keeping up their endless crawl up and down, and leaving behind them ripples of water that jostled and sparked in the sunlight. You'd have to have the soul of a grocer, not to be able to appreciate the beauty of it all.

A phenomenon of our sure knowledge that all of us were only living on borrowed time was our heightened awareness of everything; a certain wistfulness, too, that up to now, and for all these years, we had never noticed the daily wonders of life. Things like being able to sleep at night, and awaken — still alive — the next day. Things like the beauty of a full moon on a calm sea. In whichever direction you looked out to sea, the darned old moon would be laying a shiny silver path across the water direct from your feet to the far horizon.

And the dawn! We saw a lot of dawns, whether we wanted to or not. There's lots of ideal ways of seeing the dawn break the sky. It could be from the top of a faery mountain, through the high window of a turreted castle, or from the lesser thing of a casement window of a country cottage and with the birds all around on the overhanging trees, busting out their little hearts in the melodious songs they were taught, and have never forgotten oh! — hundreds of thousands of years ago.

The pre-dawn noise of the chattering sparrow, the first to dare to break the holy hush left by the nightingale, as fast-dying dark night enfolds his last melodious song; the shy cheep of chaffinch; the challenging call of the mistle-thrush, his sharp notes cutting the cool like a whip-lash; the complaining cooing of the wood-pigeon and the never-ending love-talk of the doves.

The woodpecker taps discreetly on the hollow elm tree — to waken the Sun God from slumber. 'Time to get up, Time to get up,' hammers the woodpecker impatiently. 'Time to get up and show thy face.' And if you've ever heard a dawn break in the country, you'll all be familiar with the sudden quietening of their voices just before the sun shows its face, and you'll have heard them burst into even louder song as light streams o'er the world once more. In the meadow just beyond the woods the skylark at the first hint of dawn ascends vertically skywards, his trilling hymn of praise piercing the fast-folding night. He hovers there and from his high vantage point, determined to be the very first to see the sun peep over the horizon.

Only the birds have still remembered their traditional homage as they await the arrival of the great Sun God. Only the birds begin their laudatory anthem of praise and welcome before he arrives in full splendour. Only the birds... and the hawking, coughing, spitting soldiery wakened from fitful sleep an hour before his arrival. And their greeting is not the joyous song of praise but a loud clatter of steel upon steel as they 'fix bayonets'... and wait.

The most poignant way to greet the dawn is at the compulsory 'Stand-to' readiness every day one hour before dawn. And when the sleep-sodden, weary unshaven, unkempt and bloodshot-eyed soldiers gratefully sip a tin of scalding milkless, sugarless tea while they await the first glimmer of light to brighten the sky. We'd stand around, quiet and hushed, hardly making a sound. *Not* that we were listening for any signs of enemy attack. The presence of the enemy lines facing us a hundred yards or so away meant not a thing. Our front trench ran due East-West across this little Peninsula, with the Turkish lines to the North of us. But it was not to the North we faced, but East to the unknown enemy—Dawn. If ever we thought about the Turks at all at such times it'd be probably when a gust of dawn wind might bring in its wake a whiff of the stench of death from the blokes rotting out there in No-Man's-Land.

This pre-dawn hush was something more overpowering, the epoch-making Cosmic Event of the birth of a new day. It overawed us. The first streak of light would arouse us out of our statuesque immobility and the day's bustle would begin. We'd unfix the murderous-looking bayonets that we'd stuck on the end of our rifles an hour before to greet the dawn. In what is traditionally and poetically the darkest hour of the night, Dawn was the dreaded enemy to come stealing up on us. But when Dawn came, you'd 'unfix bayonets'—there'd be a noisy clatter of steel and you'd greet joyfully the dread foe, Dawn, that you'd braced up and tensed up to meet an hour before.

'Sergeant, can I nip for a quick shit—me guts is bursting?' and the Sergeant's reluctant acquiescence—'Orlright, but look sharp. An' take thi bloody rifle wi' thee, in case...'

In case, Soldier, in case Dawn itself catches up with you. This mysterious enemy Dawn.

'Ee, but Ah don't know. This bloody Stand-to. Ah reckon it's a lot o' balls. Doesn't tha' think so, Sarge?'

But the Sarge is a 'Company man.' Such heresy from one of the workers horrifies him. 'Nay, it must be reet, else we shouldn't be doin' it, should we?'

'Greater faith have I not found, nay, not in all Israel' said somebody once. Or maybe the Sergeant comes from a long line of Druids. Maybe he still retains a race-memory of Stonehenge dawns!

Magically enough, when Dawn arrives, it's transformed into our beloved friend. Dawn is always enthralling, but never more magical than in the Med.

> Lo! the Dawn. Lo! in the East
> Flame the first fires of beauteous day, poured forth
> Through fleeting folds of Night's black drapery.
> High in the widening blue the herald-star
> Fades to pale silver as there shoots
> Brighter and brighter bars of rosy gleam
> Across the sky.
> Far off the shadowy hills
> See the great Sun, before the world's aware
> And don their crowns of crimson; flower by flower
> Feel the warm breath of Morn, begin to unfold their tender lids
> O'er the spangled grass
> Sweeps the swift footsteps of lovely light,
> Turning the tears of Night to joyous gems.

'It's bloody lovely, here, mate, ain't it? Y'know — when the sun comes up. All silver and gold like, ain't it? Bloody lovely, it is. Makes yer feel 'appy like, don't it, mate?' Well — merely a different way of saying the same thing, I guess.

Springtime and early Summer in the Med. is great — but especially on this little strip of the Peninsula of Gallipoli. All around, except where the trenches had scarred the terrain, there was a rash of spring flowers to remind you of what would be happening back home at this time of the year. Especially those little blue flowers that grew there in such profusion — never could find out the name of these flowers — but a bit like our English bluebells.

This was the time of the year back home when the kids all over the country would be carrying home huge untidy armfuls of bluebells, those heavenly flowers that oft, me thinks, Nature made especially for children. The maddeningly sweet smell of those flowers is something you never forget. They'd be stowing these flowers, the kids, in jam jars all over the house. These blue jewels, robbed with such impunity from the green carpet of England, would be in every room — on the tables, on the kitchen dressers and crowding every window-sill of the house, and you'd come downstairs in the morning to be assailed with a wave of fragrance that, if you're a bit of a nut like me, floods you with joy. In my own particular part of the world the kids had a custom, too, of paying ceremonial visits to the local cemetery to put a jar of bluebells on the graves of their grandmothers, an old local superstition that if you did this you were sure of a seat in Heaven sometime. Occasionally the kids cheated a bit; if they didn't happen to have a grandmother of their own in the cemetery, they'd 'adopt' one, stick a jar of these blessed fragrances on the grave of someone else's grandmother. Grandmother identification was not of first importance; what was important was booking your seat in Heaven!

We were nattering about these things and home during a lust of 'Home Thoughts from Abroad' when the urge came to climb out of the trench and pick a few of these local blue flowers. Some blokes never grow up — especially nuts like me and my mates. We got quite a sizeable bunch, too, of these flowers. Then some damned Turk sprayed a lot of machine-gun bullets around us. The bullets sprayed all around us in a fairly tight circle, but leaving us practically unscathed. If you know 'owt' at all about machine-gunning, you'll know it's a thousand-to-one chance against this being accidental. I prefer to think that the machine-gunner was some middle-aged Turk with lads of his own at the front, getting a quiet chuckle at putting the wind up us, but careful not to hit us, and following in a practical sort of way the precepts of his own religion — 'Allah, the all-Merciful, the all-Compassionate' sort of stuff, tha knows! Maybe I kid myself. I dunno. But Johnny Turk was not only a stout fighter, he was always a chivalrous one. And the Mohammedan, by the very tenets of his faith, is forever mindful of his obligations towards 'Allah's Afflicted', which included silly young sods like us.

Except for one of our chaps getting a slight flesh wound in the arm, no one was seriously hurt. No bones broken — all that was broken was the 200 yards speed record as we dodged back to the shelter of the trench. The moral of this incident and there's got to be a moral hasn't there? — the moral is that boy soldiers like us are always daft young bastards!

What for all this digression? I'm damned if I know, except that it's a relief from the eternal sordid surrounding picture of mutilating death, the scent and memory of these flowers transporting us on its magic carpet away from all this. This place could be a paradise if we weren't all so busy trying to knock hell out of one another. Balmy sea breezes, light morning mists that disperse quickly in the hot sun and unfolding to the view the magic beauty of the countryside. 'Just like England on a hot summer's day. Except for the bloody Turks,' said one of our chaps despondently. 'A good place for growing tobacco,' says Abdul the Turk, 'I could be happy here with a little of this land, but for those damned British — may Allah the all-Seeing, the all-Knowing, the all-Compassionate use their guts for garters.'

Me, I really loved the place. The early morning was usually quiet. Maybe the Turks had a 'thing' about starting any rough stuff before breakfast. 'Or maybe they're having an early morning session on their prayer mats, the heathen bastards,' said the young Irish boy of our Company with a line of blasphemy all his own that was a constant source of envy to us. A devout Catholic, he crosses himself reverently, 'Blessed be God, and Blessed be His Holy Name — may he strike these bloody heathens dead,' he says, combining due reverence and ill-wishing with native Irish ingenuity.

Seawards — there's the view of the minesweepers — the only sign of life on the water, the slow crawl — up and down the still blue sea — of these little boats, assiduous in their search for Turkish mines. It all looks so peaceful, these minesweepers, watching 'em move lazily along. One of our gang had a brother in the Navy and he explained the technical workings of these deadly mines. 'They're sort of round things,' he expounded, 'summat like a round dust-bin, and wi' little horns wot sticks out of 'em. And when they 'its the bottom of the ship, they blows you to bloody hell,' he said cheerfully.

The lone monitor with its 14-inch gun sends an occasional outsize shell to go slowly droning over our heads, over to the Turkish forts at Kum Kale, and we'd wait, and you could count almost up to ten after seeing the flash of the explosion and before the noise itself reached us. Our usual comment, uttered like a prayer, invariably accompanied the noise of these explosions — 'Hold that, you bastards.' Without a trace of malice in our hearts, we would mouth the incantation like a ritual response, as if it had been the response in a religious ceremony.

The sweet smell of little wood fires being lighted to brew the little mess tin of tea to start the day. The sun climbs higher and hotter and there is a hum and a song in the air — the song of summer. Shirts come off to expose sun-tanned torsos. The pre-natural quiet of the early day tricks us into an illusion of safety and we climb out of the trenches to lark about a bit and for exercise. A spray of shrapnel materialised magically and malignantly as the shell bursts with a wicked snarl stove our heads, and we know that playtime is over for the day and dodge to the security of the trench. The now old-familiar cry 'Stretcher-bearers!' is taken up along the line and yet another of these larking young soldiers is carried off to the nearest First Aid Post, or to a place where First Aid is quite superfluous.

'But for those bloody Infidels out there,' says the Platoon Officer, shaving with difficulty and finding the polished base of his mess tin lid quite distorting as a shaving mirror, 'but for those bastards, a man could have a damn good holiday here. Grand place it'd be for a honeymoon, don't you think?' and wonders privately why his fiancée back in England doesn't reply to his letters properly, instead of keep harping in her own letters recently about Jack Barclay. 'Ever so smart he looks in his Captain's uniform and all those gold tabs, now he's been put on the Staff at the War office,' she says.

'Damn and blast Barclay — bloody War Office wallah, and damn all women, too. Do some of those armchair warriors at the War Office a bit of good, to spend a week-end out here,' he mutters.

'Eh?'

'Oh! nothing. I was just thinking. Bloody women.'

His friend regards him curiously, opens his mouth to say something, then changes his mind. 'Ah, why worry,' he says soothingly, 'plenty more fish in the sea,' and grins. 'Little Turkish fish, too — charming when they're young. When we get to Constantinople, if ever we do, I think I'll find myself a little Turkish virgin — about 15 or so — yashmak and all, with her big eyes peaking over the yashmak and promising me what I don't get enough of. Yeah! and maybe I'll settle down with a fez and baggy trousers and a hookah; might even get myself a harem. I could do worse, I think. Anyway, it'd be better than going back to the smoke and grime of Lancashire again.'

'You know what you are, old man?'

'Yes, I'm Lt Johnson, handsomest man on the Peninsula, the darling of all the little bints in Cairo.'

'You're just a lecherous and lustful old bastard.'

'Yes, I am,' agrees Johnson complacently, smirking and grinning like a satyr about to set off for a dirty week-end.

'But for those cross-eyed Christian dogs, may Allah look sideways on them — but for those unholy worshippers of the Cross, I could be back now in my little village in the hills and aiding my old father in the fields. It's going to be a good year for the crops, the ancient ones say, and he could do with my help now that his joints begin to creak and his legs getting stiff. Ai! Ai! — but what would I give just to be back. And in the evening and the labour in the fields ended for the day, I would bathe and put on a clean shirt and go and sit with my little Yasmini, may Allah the all-Merciful, the all-Compassionate protect her from the Evil One. And we'd sit there in the house of her mother, opposite one another at the ends of the table and not talking a word because of her mother sat between us. But our feet would meet under the table and they'd talk together and embrace and entwine together and her eyes would promise me all the delights of Paradise once we are married. Bismallah! But how I wish that Allah would smite these Christian dogs with a poisoned dagger. Blessed be Allah, the all-Merciful, the all-Compassionate, may he crucify these Christian dogs on their own idolatrous Cross.'

Thus Abdul relieves his feelings, and bitterly. Alongside him in the trench, Mustapha listens, opens his mouth to say something, but wisely changes his mind. 'Best let him find out for himself,' he thinks. 'Too hot-tempered and too quick with his dagger, my little Abdul. Let someone else tell him, or let him find out for himself. As Allah wills.'

But in the last seven days leave that he, Mustapha, had had to help his own father plant out the new season's tobacco seedlings, it was whispered discreetly in all the little cafes in the village. 'Eating too much overripe melon,' they excused it indulgently, this swollen belly of little Yasmini. Some blamed one from a neighbouring village, others said it was one of those German officers now coming into Turkey in such numbers to aid us in the fight against these British. May Allah, the all-Merciful, the all-Compassionate, throw the genitals of these British soldiers to the dogs.

And Mustapha allows his eye to travel speculatively on his friend Abdul, and muses further. 'But Yasmini's a clever little girl. If only she can get delivered of the little one before my little Abdul goes home on leave and take the fatherless one to one of her many cousins who live in the hills, then maybe we'll have a happy wedding after all, and after the marriage ceremony, and as is the custom of our country, we'll all go running behind the wedding cab with its blinds drawn tight down on Abdul and his little Yasmini, and wait for him to throw up the blind and lean out of the window and wave the little red-stained handkerchief . And then we'll all cheer like mad because Abdul has shown us he's married a virgin. For Yasmini's a clever little girl, and the trick's been done before.

'And the truth of the matter will never be spoken by the men of our village, for all men know that women the world over are as frail as are our tobacco seedlings in the first two weeks of planting. And the women of the village will never even give a hint to Abdul, for who knows, it might come to pass — and which may Allah forbid — that one of their own daughters may have to take a hurried trip to a distant cousin — what with men being what they are and the madness that comes over young girls at the full of the moon.

'Poor Abdul, and poor little Yasmini! And to think that the whole business might never have happened but for the coming of these Christian dogs — may Allah, the all-Merciful, the all-Compassionate, burn them in a fire of dung.'

CHAPTER 15

Strictly Private and Reserved

By the end of May there was a silent rumble of things to come. Nothing you could put your finger on but it was there all the same. 'Summat's up' was the general opinion. Something was afoot, there's no doubt.

By now we had left our Shangri-La on the West coast of the Peninsula and moved towards the centre of the battlefront. The usual once-weekly fatigue of carting ammo, boxes from the beach to the front line had become a daily chore. It was hot and gruelling work, as everything on this Peninsula had to be moved and carted by hand, with the soldiers as beasts of burden. By now the season had set in for long hot summer days and manhandling these heavy wooden boxes full of cartridges along the steep slopes of the gully was exhausting. 'Cached' and distributed in convenient dumps near the front line, there seemed to be enough for the whole British Army. And small ration dumps appeared in odd places. In fact, all sorts of unaccustomed things began to happen. 'Summat's up' — there's no doubt.

Our Platoon Officer in his rounds drew our attention to the back page of AB 64 — the soldier's pay book which we all carried. Like a bank passbook, this AB 64, except that there was damn all in it of any transactions. 'It's for when we get paid — a sort of record, like,' our Sergeant explained. Up to now we hadn't drawn a red cent since we landed. Not that it mattered much. There wasn't a damn thing you could buy on the Peninsula — no canteens, no YMCA huts, no stores, in fact 'no nuffin.' Towards the end of the summer we did get a pay-out, including all back credits, and at the same time a YMCA or Church Army tent did in fact materialise and was soon swamped by customers and sold out. Shortly after, even that folded up. I don't think the YMCA blokes 'reckoned' this Peninsula much. Like the Arabs, they 'folded their tents and silently stole away.' Only quicker.

So we dug out our AB 64's from our pockets and studied them. The last page was headed 'Form of Will' and went on to say, 'I give and bequeath all that I now own and possess, both now and in the future, to... (insert name of next of kin). Signed...'

That's what the last page said, or some such similar legal jargon.

But the Officer said we ought to fill it up, 'for,' as he said, 'you never know. Accidents happen sometimes in the best regulated families, you know. And especially in war-time.'

'You can say that agin, Sir,' one of our blokes muttered and the officer grinned.

'Anyway, boys, I've been told to advise you to fill up the last page if you feel like it. It's up to you, you know.'

We got a lot of fun filling up that last page. No doubt the intention was good and it was guaranteed to be acceptable in law — which was a good thing, I suppose. Without it, think of the endless expensive litigation as the relatives of the deceased squabbled amongst themselves for their share of the untold millions these half-starved, bob-a-day soldier boys would have left behind them.

'Wot 'appens, Sarge, if y've got fuck-all to leave?' asked one of our merry irrepressibles. But the Sergeant was not to be caught. 'Well, in that case, lad,' he said solemnly, 'tha leaves it to thi next of kin.'

Some took the job seriously and filled it up correctly. Some filled in the next-of-kin as 'My dear Mother' (unnamed), and then forgot to sign the will, which must have posed many a legal headache to the lawyers. Well, it would have done if anyone had taken the slightest notice of these AB 64s. Usually on decease, the whole bag of tricks — AB 64, Form of Will, cheap cigarette lighter, photo of best girl or wife and kids, a keepsake ring or some such token — they were all lumped together and stowed somewhere. But God alone knows where!

The odd two or three hundred multi-millionaires in our battalion, well — maybe they made their own private arrangements!

But apart from this massing of little ammunition dumps and little ration dumps near the front line, apart from timely reminders anent the possible future value of the back page of the AB 64 — this ominous 'Last Will and testament' page that gave us all such good sport in completing — quite apart from all this there were other little straws in the wind, telling us that 'summat was up.' The officers started visiting us two or three times a day instead of the usual once, and chatting us up like we were blood brothers. 'Never knew an officer could be so chummy,' said one of our lads happily. 'There we was, like, talkin' 'bout bloody mill back 'ome, 'im and me, like, as if we was 'piecing-up' together. And 'im askin' me all about me brothers and sisters — just like old pals we was. Told me, too, 'e did, as 'ow 'e was going back to his feyther's mill after this bloody lot's over.' (The fact that this young officer's ambition was never realised was due to circumstances quite beyond his control — shrapnel being no respecter of persons).

Unknown faces appeared in our trenches, carrying lists of names, seeking out chaps by name, unknown faces in Officers' uniform and wearing collars that fastened at the back instead of at the front, the ecclesiastical crows that foregather in spiritual converse with their lesser brethren. And as if making up for hitherto neglect, little impromptu prayer meetings and cosy spiritual chats were held in odd comers of our trench. Each to his own, these visitors sought out erstwhile neglected members of their flock — Methodists, Congregationalists, Presbyterians, Plymouth Brethren, Unitarians, etc., etc., so these various priests of the many cults of our great religion administered spiritual comfort and consolation, and all in a spirit and manner of jolly good fellowship. In fact, you never saw such display of contorting good fellowship as exuded from those 'dog-collar' officers.

'I wonder if they'll be with us when the battle starts,' one our blokes mused aloud. His pal spat disgustedly. 'Don't be so bloody daft, mate. Where do you think they'll be.'

Anyhow, they made a grand show of earning their living two or three days before the guns opened up. 'Never knew there were so many bloody parsons about,' said one of our chaps wonderingly, forgetting that the ODs (Odds and Sods) embraced a multifarious collection of the various sects of our Christianity.

'In my Father's house there are many mansions,' said Jesus, and implying that there was room for us all. But when you come to reckon up the numerous and ever-increasing varieties of differing Christian cults, there'll be a real need for many mansions. There could even be a housing shortage!

From a certain vantage point in our trench, you could get a glimpse in the distance two or three times a day of a white-surpliced C of E chaplain, amidst a gathering of the more numerous Church of England soldiers. 'Anyone as wants to attend Divine Service is excused fatigues,' said our Sergeant magnanimously. We had been used to seeing these voluntary attendances at services on odd Sundays, but it seemed a little weird to be having them on week-days. Strictly one-day-a-week religionists, us, we found it a bit strange, and slightly obscene — these week-day services. If there was to be a big battle to come off any day now, why drag religion into the dirty business. I guess most of us would have preferred it to be strictly a soldier's job, and not to sanctify it in the name of religion. Some of our blokes attended these services, a bit apprehensive of the future. Some of us set off to attend, but sought out a crafty hiding place for a quiet snooze, sleep being always in short supply.

'And we ask, O Lord, that Thou wilt steel these Thy servants' hearts in battle, and grant that they may be victorious in their righteous struggle against all evil towards our Gracious Sovereign King George V, and in defence of our worthy and just cause against the Infidel.'

In a little corner of the trench nearby where I was doing a spot of sentry-go, some such similar words as these floated up to me, sonorous, impressive, and oddly sincere, and the three or four blokes there with bowed heads whispered 'Amen.'

(Meanwhile, God, in His occasional pseudonym of Allah, is equally being implored by the Infidel Turks to beat the living daylights out of these damned British. 'What *am* I to do?' He says distractedly, 'I can't please both sides.' Maybe He spins a coin.)

'Sustain them O Lord when sore distressed and wounded in battle, and comfort them.' (Irreverently, I thought, I'd rather settle for a shot of morphia). 'And finally O Lord, receive them into Thy Kingdom that Thy saints may acclaim in one voice, "Well done, thou good and faithful servants; enter thou into the joy of Thy Lord."'

The Medieval liturgic phrases had an illogically comforting sound, reminiscent of the quiet backwaters of Church services heard as children.

CHAPTER 15

'Washed in the Blood of the Lamb, we shall be whiter than snow.' The uniformed and most reverend gentleman drones on.

'Wash me in the water where you wash your dirty daughter, and I shall be whiter than the whitewash on the wall.' The words of one of our ribald marching songs echoed mockingly in my head. The similarity was painful, and the lines continued to echo mockingly as I listened to the parson's words. Then there was a spot of heavy shelling for a while and the parson hurried through the Lord's Prayer, promised to come again tomorrow, then nipped back smartly to the more secure regions of his base in the lee of the side of the gully.

True to his promise, the parson — one of the OD Sects — returned the next day, the day before the battle. One of our chaps, a recent reinforcement on the last draft, had missed the last of the C of E Services, but hearing that a parson was conducting a small service in our sector of the trench, came along happily, his face well-scrubbed and clean, and his uniform neat and proper. In his hand he clutched a nice Prayer Book of soft vellum, not one of the cheap books issued to us on enlistment. There was a calm and happy look on the lad's face as he passed us on his way to the Service. A decent sort of youngster this, and not like us to whom cuss words were the very breath of life. We didn't even swear when this kid was around. Well, not so much anyway. One kinda pictures him having but recently left a quiet God-fearing home in one of the more respectable streets of my home town.

So, if he was happy, I guess we were just as happy for him, too. Maybe some of the inner goodness, some of the clean piety of this kid dropped on us accidentally. 'Tha needn't be in a hurry to get back, lad,' one of our chaps said kindly as the lad passed. 'Ah'll draw thi rations for thee.'

'Stop thi bloody swearing,' our Sergeant muttered to one of our gang, busy relieving his feelings while de-lousing his pest-ridden shirt.

I do believe we all secretly shared the hidden joy of this youngster on his way to an appointment with spiritual consolation. But soldiers are like that — at least, they are on Active Service. Revelling in the primordial earthiness of their daily life, but always their eyes hopefully following any little gleam of light that may, as the old hymn says 'point the way to the skies.' We'd have been a push-over for Billy Graham!

The lad passed us, his face calmly happy, his eyes glowing gently.

Dammit — you could almost see the halo round his head. 'Nice lad, that' says our Sergeant gruffly, as the lad disappeared behind the traverse.

We were quite unprepared for his sudden re-appearance some two or three minutes afterwards. Stony-eyed and set-faced, he re-passed us without a word and went back to his own sector of the trench. 'Wonder wot the bloody 'ell 'appened theer,' said our astonished Sergeant.

We learned later from one of the blokes who'd actually been to this tiny service. 'Well, it was like this 'ere. This lad, see, comes up, all friendly like and 'appy, and our parson shakes 'ands and they 'as a short chat, like, just afore 'e begins 'is little service. Then one of our blokes, an' not meanin' no 'arm, says to the lad, "Ah thought you was C of E, mate." "Yes, I am," says the lad, "but it's all the same isn't it?"

'But parson was shocked and said 'e couldn't get givin' a service to C of Es because the C of E chaplain wouldn't like it, like. He was very nice about it, was this 'ere parson, all quiet and gentle like with the lad and sorta quietly shoos 'im off, like, and we starts to get on with the service. Ah s'pose this 'ere religion lark is a bit like a business — they can't pinch somebody else's customers, can they?'

'Then the lad says, "But there ain't going to be no more C of E Services, Sir, as the last one was this afternoon" and as 'ow 'e was on fatigue duty on the beach and couldn't get to it an' wot with the battle starting tomorrow, 'e thought, like..."

'But parson says 'e's very sorry and says summat 'bout it not bein' etiquette or summat. Buggered if Ah knows wot 'e meant, anyway, 'e wouldn't let 'im come to t'service. Said it wur a private service reserved only for those of the same sec'.'

Possibly the parson had good reason for with-holding his spiritual comfort. I dunno. Maybe what the Good Book says about 'Suffer little children to come unto Me' didn't include boy soldiers. It's all very puzzling to laymen like us.

The older I get, the more I ponder on what a real 'dog's dinner' we seem to have made out of the simple teachings of the Carpenter. Where have we gone wrong? This monstrous conglomeration of sectarian taboos, church rituals and superstitions, enslaving encyclicals, absurd ecclesiastical laws and observances and the whole damned nonsense wrapped up in a cloud of pseudo-mysticism that would make a West

African witch-doctor green with envy. One thing's for sure, if ever the Carpenter took a notion to re-visit this Earth, He'd have one hell of a job to recognise His own religion.

Some days later when the roll was called, this lad's name was answered by the now monotonous call of 'Missing.'

'Missing' on a roll-call parade meant 'Damned if I know where he is — probably killed, but we can't be certain sure till we've cut off the string round his neck to get his identity discs.'

Discussing casualties later that day, this lad's name cropped up. Irreverent lot of young heathens that we were, we felt strangely moved that the one decent kid in our midst should have been denied some spiritual consolation before being called upon to meet his new Celestial Commanding Officer, and our Sergeant himself seeming to be more sad than any of us.

We had in our gang one Fred Buckley whose sly wit was a constant joy to us.[13] He was for ever chuckling to himself as if at some private joke, and while we sat there in the trench with glum faces, and thinking of the nice little lad who had now left us for good, Buckley's own face was creased in grins.

'Wot's so bloody funny about it, then,' said the Sergeant sourly.

'Oh! nowt special, Sarge. Ah wur just thinkin', like.'

'Wot about?' said the Sergeant suspiciously.

'Oh! nowt special. But tha knows, Sarge, when we used to line up for Church Parade in Cairo, and afore we marched into t'C of E Church, t'owd Sergeant Major would always give the command "Fall out the Jews and Roman Catholics".'

'Well, wot's that got to do wi' it?'

Buckley spluttered with laughter. 'Oh! Ah wur just thinkin' Sarge; maybe Heaven's a bit like that. Reserved for a special Sec', like.'

The merriment subsided and we became glum again, thinking of this pious little lad. Says the Sergeant wistfully — 'Ee, Ah dunno. All these stupid different Sects and religions. Pity that parson was such a stupid bugger — and a nice young lad like that n' all. An' wot difference

---

13 Probably Pte 10190 (240949) Fred William Buckley. See Appendix II, Notes on individuals mentioned in the text.

would it a' made?' he asked angrily of a nearby Corporal. 'It's all the same bloody Firm, ain't it?'

We all joined in vociferous denunciation of the erring Shepherd, until the Sergeant told us to shut up. 'Discipline's discipline,' he said, 'and parsons is rated as officers. So pipe down, all of yer.'

* * *

An unforgettable character, this bloke Buckley. The only time anyone ever saw him angry was at the medical sort-out in Cairo when we were getting ready to go to Gallipoli. Buckley had been rejected because of his malformed trigger finger — result of an accident in a saw mill as a boy. Inadequate and primitive surgery had left his right forefinger grown out at right angles from the second joint, but in spite of this, Buckley, like most of us, had gained his extra 3d a day as a first class shot. He stormed and raved at the rejection, and even button-holed our own CO, Lord Rochdale who, accompanied by the RSM, was crossing the Citadel Barrack Square.

Omitting even to salute he angrily demanded of the Commanding Officer why he had been taken off the list of those due to sail to Gallipoli in a couple of days' time. Ignoring the RSM's stern admonishment 'Stand to attention when you speak to an officer,' he dismissed the outraged RSM with a contemptuous 'Thee shut up — Ah'm talkin' to t'Commanding Officer' — and pleaded his case so belligerently that a stunned CO at last agreed, and Buckley resumed his old chuckling happiness and left the CO after a posh salute and a heel-click you could have heard half a mile away.

'Leave him to me, Sir,' said the furious RSM, 'I'll sort him out. I'll teach him to approach an officer like that.' But the CO's gaze was directed at the back of the retreating figure. He grinned. 'If the whole of the British Army's like that young devil, we can't lose can we, Sergeant Major? But give him 7 days CB. We'll be leaving, anyway, in a day or two, so he won't do them, but he's got to be taught discipline, hasn't he, Sergeant Major?' And even the Sergeant Major smiled, contenting himself with a muttered, 'The saucy young bugger.'

Once during a temporary spell of fatigue duty on Gallipoli in our improvised cock-eyed, open-air cook-house about 200 yards behind the front line, Buckley would often leave his cook-house duties and find himself with us in the front trench. His visits always coincided with some tit-bit he had brought for somebody — a bit of fried steak he'd nicked and brought up to someone of us who looked extra peaky and under the weather, or a tin of milk, or some extra sugar. It's bloody marvellous how little kindnesses like this count. And when you're feeling extra depressed and low-spirited, hating these days of mutual callous suffering, that feeling you get sometimes in the front line, that feeling that nobody don't give a monkey's f— no more what happens to you... or to anybody. Then someone like Buckley comes along, with his harsh Todmorden and barely understandable dialect and his broad grin, and quietly slips you a nice lump of cold juicy fried steak that he'd pinched from the cook-house, and with it a few words of Rabelaisian comfort. "Ere, get this inside thee, Charlie boy. An' then tha'll feel like tha could do wi' a good woman again.' So you mentally bless this character with the grotesque trigger-finger joint. You get that comforting feeling that there is, after all, someone left in the world who's got his eye on you — someone quietly, who cares whether you live or die. Then, all of a sudden, life seems better.

He seemed for ever to be on the look-out for someone he could give a bit of practical help to. And his snide remarks and caustic wit was a constant tonic to all of us. One day, and while temporarily deserting his cook-house duties, he was with us in the trench when a hand-grenade attack suddenly materialised from Johnny Turk. Instead of nipping back to his proper duties in the cook-house, Buckley joined gleefully in the fray until an exploding grenade tore out his left shoulder. I never knew what happened to him after he was carted away, groaning, on an improvised stretcher, or whether he lived or died. For me he still lives.

CHAPTER 16

The Pot Boils

In the first week of June, June 6th if my memory's correct, the pot boiled over.[14] The battle started in a way that was pure joy to a war schizophrenic like me. I label myself thus — more in ignorance of the proper term than in self-deprecation. But what else could you call me? One part of me glorying in the splendour of it all, and the other part... that little black goblin of fear and apprehension that every soldier keeps hidden at the bottom of his garden.

At 10 a.m. by the synchronised watches, the crescent-shaped line of the light-grey painted ships of the Fleet ceased their quiet languid basking in the sun. They'd been lying there dead still on this calm Mediterranean lake ever since the sun had dispersed the light morning mists and exposed them to view. Idle, lazy, painted ships upon a painted ocean. They were so dead still, you'd think you were looking at a framed canvas in oils. But for the occasional wink and flash in the sunlight as the ships' brass caught the sun's rays, it was like looking at a masterpiece by some naval artist. But almost to the second, at 10 a.m., all hell was let loose and the ships became a searing line of flash and flame. This naval bombardment was the preliminary to the battle, our battle, in which we were to survive — or not. As Allah wills! Twenty minutes of this, we were told, then 'over the top.'

Waiting for the minute hands to crawl round to 10 a.m. and for the bombardment to begin, we appraised this line of ships behind us. From

---

14 Watkins' memory was not correct! The Third Battle of Krithia began on 4 June 1915.

our vantage point on the higher ground, they lay there on the calm sea, like ships at anchor at a naval Review. You sort of felt that something was missing. They should have been 'dressed-over-all' — or whatever it is these naval blokes call it — y'know what I mean — strings of flags and bunting all over them from mast to mast, and the Royal Yacht steaming slowly along the line. And all the matelots hand-in-hand lining the rails, faces scrubbed clean, uniforms spotless, hat bands gleaming white, and the music of the ships' bands adding to the gaiety and splendour of the scene. And on the shore, there should be a joyous crowd of sightseers, glorying in the British Navy. The Power and the Glory! Like the 1910 Review at Spithead.

On our own shore we had an equally appreciative crowd of sightseers, equally appreciative albeit a mite apprehensive. Damn and blast those little black goblins at the bottom of the garden, spoiling what should have been a perfect day out. Sun shining brilliantly, sea sparkling and clear like a day at Blackpool. The brass work on the ships continued to wink and blink in the sunlight. The pairs of little minesweepers continued their long leisurely crawl along the waters. As peaceful and as gorgeous a summer's day as you could wish to see. We should be reaching now for our bottles of pop and ginger beer, and bringing the sandwiches out, and waiting for the ice-cream man to come along — instead of taking sly sips of our rum ration to bloody well calm the nerves.

With a roar and a bang like the end of the world the big guns of the ships opened up precisely at 10 a.m. and the shells screamed over our heads, sousing the enemy ground till the whole place shook like an earthquake. The light quick-firing guns of the smaller ships kept up a continuous firework display of flash and smoke as their lighter quick-firing guns roamed and searched the smaller enemy targets. Less frequently the 15-inch guns of the flagship let off a broadside that rattled your very teeth. A naval bombardment is a grand sight, though I'd hate to be at the receiving end of those guns.

Somewhere, someone must exist, denied my pitiful inadequacy of words, who could properly describe the intense elation of this 'music and the thunder of the guns.' And even discounting the fact that I really am a bit of a 'nut,' there's something weird and fascinating about this bloody Peninsula. The feel of legendary history was tangible and real and you sensed all around you the crowds of all the warriors of all the long past wars, pushing and jostling for a ringside seat at this bombardment. I'm bloody sure this place is haunted.

We watched and watched. Not for the first time in our history did the soldiers feel a profound gratitude for the British Navy. I labelled myself a war schizophrenic, but I'll bet the most ardent pacifist of our race would have wanted to reach for his sword and lead a death-or-glory charge, once he'd had a sight of a naval bombardment.

What the hell is it about battle that is at once so scarifying and so ennobling. Is it that men are proud to feel that at long last they have been picked to undergo the final test, their gratitude 'Now God be thanked, He's matched us with this hour.' And even should he fail the test, he's proud he went through it. The Moment of Truth is a solemn moment. Where from above gleams the searchlight on the soul. If we survive the test to our satisfaction, what matters all else? What matters, if back in civilian life we are failures, or plumb the depths of the degradations of vice and drunkenness? All these are mere straws compared with the fact that we've undergone the final test — and won, or lost — according as we're made.

The hour creeps to 10.20 and we brace ourselves. 'There can't be a bloody Turk left alive after that lot,' one of our chaps said, in awe at the havoc and devastation in front of us. A perfectly natural mistake for any soldier to make, as we were to find out in a few minutes.

Myself, I wasn't to go over the top this time. Our machine-gun section had been detailed to remain in the front line and cover the advance on a strictly restricted arc of fire. I can't say I was sorry. Any soldier who tells you he is sorry is either a psychiatric case, or just a plain bloody liar.

The guns of the ships ceased simultaneously and the momentary contrasting silence cut the air like a whiplash. This was the signal, and from our trench a sea of bayonets disappeared into the clouds of choking dust and smoke of No-Man's-Land. Instantaneous impressions register on the mind like a photograph, indelibly and accurately. This was the first time I was not to accompany my mates, and it gave me the opportunity to see my own self reflected in them. Desperation struggling with apprehension, apprehension giving way in some cases to stark terror, exultation lifting some of them above craven fear — a bit of all these things I saw in the faces of these chaps as a bit of myself in similar circumstances. But to a single man, every man-jack of them went over, many for the last time.

> Into the jaws of Death, into the mouth of Hell,
> Rode the six hundred.

The British Army doesn't change much. This time it was only the horses that were missing.

By the sheerest bad luck our machine gun section was unable to give our lads even the slightest support in covering fire. Just when a stream of machine gun bullets should have been making some of the enemy keep their heads down, our machine gun team had been temporarily paralysed by a shell from one of the ships falling a bit short, just at the close of the bombardment. None of us received even so much of a scratch, but the proximity of the explosion was such that our bodies were numbed and paralysed. A sort of temporary shell-shock, I guess. Strain as we would, we couldn't make our quivering bodies obey our wills — it was like as if an anaesthetic had been pumped into our spines. It was a good three or four minutes before we were able to master our frail flesh and move our arms and legs again. 'I couldn't 'a moved a bloody inch, mates, no — not if you'd given me a five pun note,' one of our chaps described it. We felt real bad about it — not being able to give the lads our support, but by mutual agreement we decided to keep quiet about our unavoidable defaulting. In battle, your mates never like to know they have been let down — no matter what the reason, and with the Lancashire lads the code of mutual aid is more rigid than the Laws of the Medes and Persians. 'Appen Ah'll tell 'em wot 'appened, after this bloody war's over, and when we're 'avin' a nice pint together back home at the 'Swan.' Maybe they wouldn't be so bloody peeved as if Ah told 'em now,' said our Machine Gun Sergeant. He never did get round to telling them — never got a chance to, in fact. Months later, two or three days before the evacuation in fact, the Turks were making a last desperate attempt to break through our front line.[15] They were unlucky, but it was

---

15 The diversionary actions at Helles on 19 December 1915. See Appendix III, Notes on events and places mentioned in the text.

fierce while it lasted. The Machine Gun Sergeant was all het up and agitated. I was struggling to free a jammed belt in the machine gun.

'Can tha manage on thi own, Cobber?'

'Ay, I can manage, Sarge. Why, what's up?'

'Well, the lads is a bit pushed, like. Hundreds of the bastards trying to break through our line. Me and the Corporal thought as 'ow we'd go and lend 'em a 'and, like. Sure tha can manage?'

'Ay, I'll be all right. Good luck, Sarge.'

'Ay... well...' He paused, regarding me doubtfully. Eventually I hit on the right cuss words and got the belt free. Then the fusillade of shots grew louder and he and the Corporal picked up their rifles and dodged up the little sap leading from our gun emplacement to the front line to 'lend a bit of a 'and, like.' The Sergeant was a short chap, about 5′ 2″, and the Corporal a long lanky 6′ 3″. But the difference in height didn't seem all that noticeable when they were laid head to foot on the trench floor a few minutes later.

Then we lifted them up and arranged them tidily on the firing-step. One of the lads tucked a blanket reverently around each body, retching slightly when he came to the shrapnel-mutilated red pulpy mass that had once been the head of our well-loved Machine Gun Sergeant.

'And to think that only another day or two, and they'd 'a bin off this bloody place for good,' he said. 'Well, they're off for good all right, now,' his mate replied, and ending with the soldier's usual obituary, 'Poor bastards.'[16]

But in these obituaries, it's not the words that count so much.

---

16 The sergeant is probably Sgt 7646 John L. Taylor Mellor. There are five candidates for the corporal. See Appendix II, Notes on individuals mentioned in the text.

CHAPTER 17

The Philosopher

Drafts from the Reserve Battalion back home continued to arrive straight off the ship on which they had embarked at Southampton, and landing on the Peninsula even more ignorant of war than we had been. They came straight up to the front line from the beach to fill the ever increasing gaps in our ranks from previous casualties — and in turn being replaced themselves as they, too, made way for others in this endless stream of war expendables. Maybe it's our turn next — quien sabe?

Who knows, indeed! Anyone of us today, or the next day, might enrich this foreign soil, to make yet another

> ... corner of a foreign field
> That is forever England.
>
> In that rich earth a richer dust concealed,
> A dust whom England bore, shaped, made aware,
> Gave once her flowers to love, her ways to roam,
> A body of England's breathing English air,
> Washed by the rivers, blest by suns of home.

Oh! damn! Why can't I think of things like that to say, instead of drooling away like an old soldier after the sixth pint.

The sense of ominous Fate was ever forebodingly with us. Our turn next, but when? Our turn to take our place in the long queue to cross the Styx where Charon waits fretfully on his oars.

And we who now make merry in the room
They left, and summer dresses in new bloom
Ourselves must we beneath the couch of Earth
Descend, ourselves to make a Couch — for whom?

and nowhere in the world is this more ever present than in war.

These lads from the Reserve Battalion were doubly welcome — not only to fill our depleted ranks, but they were all bearers of the latest news from home. A territorial battalion such as ours, all from the same town, we all knew each other fairly well. The latest gossip, the juiciest scandal, good tidings, bad tidings, all the lot these new arrivals brought with them. A lovely nostalgic air hung around these reinforcements. Blokes would even give them half their sacred little store of rum hoarded so miserly drop by drop against an emergency.

Ah! Great is the news from home. No wonder we laid down the red carpet for them. And most of them had landed off the ship with well-filled cigarette cases, and they generously handed them round.

Cor! The first taste of a decent fag after you've been nearly crawling up the wall for one. Though anyone of us would have nearly committed murder to get a packet of fags, none of us would ever accept one without first warning the careless giver that there was a fag famine in the land and that once his fags were gone, he'd get no more.

The conscience of a soldier! He'd damn near cut your throat to get your fags, but never accept one as a gift unless the giver had been warned of the famine. Methinks that to thoroughly understand the conscience of a soldier, you'd need a blueprint, or maybe a working model of it and will full explanatory notes!

The newcomers were dispersed along the trench individually so in each sector we had one to ourselves — the one to ply with questions, one to retail to us all the latest happenings since we'd left old England's shores in August the previous year. It was only necessary in our trench, any day, to hear rumours of a new draft arriving. That's enough to send us all twittering with excitement.

You could always spot them, these new drafts. They stood out like a sore thumb. Neatly dressed in clean regulation uniform, trousers properly creased, cap peaks stiff and clean — 'they've even got their bloody buttons polished' one of our blokes commented disgustedly. Infamy, it seemed, could go no deeper! Real soldiers they looked like. We would glance uncomfortably at these reminders of peace-time soldiering.

Equally they were non-plussed at our appearance. Putting it mildly we were a trifle bizarre. 'Wot's up, Jack,' one of the newcomers greeted an old friend, 'joined the Navy?' and burst out laughing. Enough to make him, I guess. Jack was sporting a lovely two months growth of beard and his headgear — a sailor's round hat. Like the bloke on the Player's 'Navy Cut' tobacco advertisement, Jack looked. The hat he'd filched from the body of one of the Royal Naval Division boys who had landed with us in the precarious early days of the campaign. Quite a lot of these Navy boys got killed early on. It seemed damned queer at the time in those first few days, alongside us in the trenches, navy blue uniforms, naval high boots and round naval hats, alongside our scruffy khaki uniforms. Hard as nails these navy boys, who joked with death in the language of the fo'castle. Until we met these navy lads we thought we knew every cuss word in the English language, which just shows how wrong you can be, doesn't it? After the first two or three weeks after we had landed, they were withdrawn to go and fight somewhere else on the front, but they left behind them many mementoes of their short stay — things like bodies and naval blue round hats and naval leather belts, oh! — all sorts of things that we collected and treasured like magpies for a time, then forgot 'em.

In no time at all the reinforcements aped our free-and-easy deshabille. Shirt sleeves would get cut off above the elbow, they'd dent and bash their properly shaped caps till they looked like real front-line soldiers. Within days, their English cod-fish-white complexions would acquire a Mediterranean tan, giving them a new happiness.

How quickly and joyously does man revert to hobo simplicity. Given the right conditions he just can't wait to toss convention over the ashcan and to be at one again with Nature's careless pattern.

The last of their fags smoked, the last bit of chocolate eaten that they had landed with, the last bit of 'bacca finished, they merged easily with us into one vast homogenous pattern of 'have-nots' and were blissfully and newly happy.

There was a Bloke once who told us it's easier for a camel to crawl through the eye of a needle than for a rich man to get to Heaven. I don't think any man really understands this simile until he's had an enforced period of 'not having nowt,' when you're living a sort of monk's life of no possessions. So you begin to scrape the bottom of the barrel for something to share, things like companionship, understanding of your mates' points of view, sympathy with their problems, compassion for their misfortunes, and for their faults at times — oh! all the other little things that when lumped together make a workable day-to-day religion. If you've got possessions, you get distracted, peeved, jealous, envious. It's not so much, I think, that possessions in themselves are evil, just that the bloody things make us deaf to the sweeter music of life.

Coming back with a bump to our dusty, sun-drenched Gallipoli Peninsula, one of our chaps gave us in detail the key to happiness and expounded the second of the four Buddhic Noble Truths. The Buddha himself might have expressed himself more poetically, and certainly more delicately, but not more succinctly than did our George. In the homely dialect of his forefathers, and with a plethora of double negatives and a sprinkling of a little old Anglo-Saxon, he expounded his philosophy and of the joy that cometh to man by his non-attachment to things material.

'Wars ain't so bad, like,' he said. 'Except for the bloody lousy rations and the risk of getting bumped off. But some way, y're happy. And 'cos why? 'Cos y'ain't got fuck-all. And when y'ain't got fuck-all, y've got fuck-all to strive for and y're a decent bloke to yer mates, see? And yer mates is just the same, see? None of us ain't got nowt, see, so we're all equal, like — nowt to look after and to worry about keepin' safe, and nowt to fret about losin'. An' we're all dressed the same, see — so we don't 'ave to worry about puttin' on a stiff collar and a Sunday suit and all that balls, so you can look a bit smarter than yer neighbour. We're 'appy just because we ain't got fuck-all.'

This primitive reasoning seemed logical and sounded an echoing note with all of us. Basking in the joy of possessing nowt, we'd nowt to lose. The most we ever could lose would be a life, but as we all had one to lose, no-one had an unfair advantage, or possessed anything his mates didn't possess. Equality of possession reduced to refined simplicity. Life's geometrical problem solved by the familiar 'reductio ad absurdum.'

CHAPTER 17

The exponent of this startlingly new and agreeable theory continued, 'Now take when y're in a pub, like. Sometimes y're a bit bothered lest y'ain't got enough cash to stand yer round. But 'ere, yer don't 'ave to bother none whether y've not got enough cash, or whether yer mate's got more'n you. 'Cos none of us ain't got no cash to buy no beer, we don't 'ave to bother about keepin' our end up. And as we've not got no bloody beer to buy anyway, 'cos there ain't none, even if we 'ad a pocketful of brass, well — it makes life easier, if y'see wot I mean.'

A round of ironical applause greeted this unexpected outburst and the orator reddened in confusion. L/Cpl Greenwood, one of our more scholarly types, eyed quizzically the exponent of 'not never 'avin' nowt' as being the key to happiness on this earth.

'Quite a philosopher, arn't you, George?'

George bridled. 'Nay, Corporal, there's no need to be calling me names, is there?'

'I'm not calling you names, George. A philosopher is a man who... oh! skip it, but it's nothing bad, I do assure you.'

'Oh! well, that's all right then, Corp.,' said George, mollified and friendly again.

Later George collared me in a quiet corner of the trench. 'Hey, Cobber, wot's a filli... something — tha knows — wot Corporal Greenwood called me?'

'Philosopher you mean, George?'

'Ay, that's right, Cobber. Wot's a philosopher?'

I answered him, rat that I am, but the temptation was great. 'A philosopher, George, is a man wot ain't got fuck-all and likes wot 'e ain't not got.'

'Ah-h-h,' says George comprehendingly. 'Ee — an' am I one o' them — wot d'you call 'em — philosophers?' He beamed with pleasure.

'Ee, but tha' art an' all, George.' Then we both laughed.

CHAPTER 18

One Moment in Annihilation's Waste

These reinforcements — they just couldn't wait to shed the last of their Reserve Battalion prissiness, the tell-tale marks of the newcomer, the neatness, the polished buttons and the English pallor. In turn, they themselves would come to eye scornfully the next lot of newcomers, eye them with the disapproving scrutiny of front-line soldiers gone native. Within days almost, we of the original gang would absorb them, even forget completely they'd ever been reinforcements, everybody quickly merging into one homogenous mass of old campaigners.

Except the odd one or two. Like Johnny Stirk for instance,[17] and he never got a chance to bash his cap in or even to blacken his buttons. No! It's not his real name, in fact I'd never seen him before, but it was 'Johnny' something or other — I forget his last name. Just arrived in our sector of the trench, bursting with eagerness and curiosity, wouldn't even wait to divest himself of his full pack, but must needs stick his head over the parapet. 'I must take a look at these Turks.' Too late to hear the warning cry of one of our chaps, 'Get down, tha silly young bugger, get down.' Before you could count three, his forehead was neatly drilled. These Turkish Snipers are terrific. Within 60 seconds of arriving in our trench and Johnny now *non est*. But it took longer than 60 seconds to hide him out of sight. It must have been a good 20 minutes before we'd stamped him firmly down in the trench floor beneath our feet. Even our Platoon Officer wouldn't have the job of consoling the boy's mother — he never even saw the lad. Best leave it to the War Office telegram, they're

---

17 See Appendix II, Notes on individuals mentioned in the text.

marvellously non-committal. And leave it, too to some more permanent record in time, such as his name on the town's War Memorial — 'Died for King and Country.' The old saying 'curiosity killed the cat' wouldn't look so good on a War Memorial.

A few minutes after we'd disposed of Johnny and shared out what precious fags he had left in his fag case, sorrowfully rifled his pockets for mementoes of his short life on this earth, things like an odd photo or two, one or two letters, an illuminated pen-knife inscribed 'Present from Blackpool', and handed these things in to our Platoon Sergeant — a few minutes after we'd done all this and made our usual comment anent the shortness of life these days, one of Johnny's old pals from back home, one of our original gang, came seeking him. 'They tell me Johnny Stirk's here.' He was excited and bursting with eagerness at the prospect of meeting his old buddy again. There was an uncomfortable silence.

'Did tha know 'im well, lad?' asked our Sergeant.

'Did Ah know 'im? Did Ah know 'im? Why, we was 'piecers-up' together at t'Sparth Mill, Johnny and me!'

The Sergeant didn't reply, but half turned his back, whistling shrilly and fussing and fidgeting aimlessly with the bits of his kit laid out on the firing step, like a man searching for something he might have mislaid. It was one of those 'Oh, blimey, what happens now' sort of situations, all of us feeling hellish uncomfortable, and the Sergeant most of all.

But the seeker-after-Johnny persisted in his enquiries, in an eyeball-to-eyeball confrontation with the Sergeant. 'Ah'm askin' thee again, Sarge. Where's Johnny?'

The Sergeant temporised. 'Was 'e a pal o' thine?' he asked.

'Was 'e a pal o' mine? Was 'e a pal o' mine?' said the seeker-after-Johnny impatiently. 'Didn't Ah tell thee, Sarge, we was 'piecers-up' together at t'Sparth Mill afore this lot. 'Appen we'll be goin' back to t'Sparth again, too, after this lot's over.'

'Ay, tha might lad, tha might. But Johnny won't piece-up no more, Ah'm thinkin'.'

'Wot's tha mean, Sarge, won't 'piece-up' no more?'

In some of life's more poignant moments, the Lancastrian will fold himself defensively in the grim witch's cloak of cruel humour.

'Well, lad, yer could say, like, as 'ow Johnny's got the sack.' And the Sergeant sat himself down on the firing step, his hand roaming defensively his empty pockets for a non-existent cigarette butt his to ease his high embarrassment.

'Ah don't follow thee, Sarge. Where's Johnny?'

There was a bit of a silence. Then the Sergeant inverted his cotton-spinning spatula-shaped thumb floorwards.

'Theer,' he said laconically.

* * *

For quite a time after Johnny Stirk's pal had returned, shocked, to his own sector of the trench, we sat around, quiet. Nobody seemed inclined to speak. It was left to Fred Butterworth to break the clammy silence. 'As tha knows reet well, lads, Ah'm a fitter and turner at Tweedale and Smalleys' (a local machine shop). 'Ah knows nowt about yer cotton factories. But wot's a 'piecer-up' when 'e's at 'ome, like? Ah've never 'eard o' them before.'

Tom Sutcliffe enlightened him.[18] He, too, had been a 'piecer-up' before the war, at the Standard Mill, the mill that adjoined the Sparth Mill by a mere 200 yards or so. So Tom expounded. In his usual lazy leisurely way of speaking, he painted the picture, at times good-humouredly, but mostly in a bitter undertone and told Butterworth of these 'piecers-up', the young lads who toiled endlessly 10 hours a day in our cotton factories, joining together the broken wisps of cotton that the machines spewed out in endless long rows as the bobbins whirled like mad tops, taking up the strands and arranging them neatly around themselves, ready for the weaving shed when full. Quite a skilled job, this — endless lines of broken strands to be 'pieced-up.' It demands skill, quickness of eye, dexterity, but above all the knack to seize the two broken ends quickly and with a twist of thumb and finger, to 'piece up' the broken strands into one. Hundreds of these broken strands to be tended, endlessly and quickly.

Some days it's all spinning of Coarse Counts, and other days Fine Counts. It's all right when it's Coarse Counts. The threads don't break so easy. And if the factory floor humidity is right, you can go sometimes for minutes at a time without a thread breaking — and giving you time, and if the master-spinner himself wasn't looking, for a bit of a lark with the

---

18 See Appendix II, Notes on individuals mentioned in the text.

piecer-up on the nearby machines. And you can even join in the singing that goes on endlessly on the factory floor, voices raised to a special soft key that is distinctive above the song of the whirring bobbins, a special pitch of voice peculiar to factory workers on the spinning mill floor.

But if it's Fine Counts, the bloody threads seem to break even if you look at 'em, and if you cough or spit near 'em you'll see a whole shower of these bloody threads break, and the spinner'll go raving mad and cuss and swear at you. It's a frustrating job, especially if the humidity of the spinning floor drops below its minimum, and leaving the 'piecer-up' to crawl home listlessly and dejected because the master-spinner had cussed him too hard. And in those transportless days there were no beneficent buses to lift your tired body home.

'But most times a 3 mile walk,' said Sutcliffe, unburdening his soul bitterly as he re-lived those days before he joined up in the army.

'Not so bad in the summer, but in the winter, and wi' the bloody snow seepin' down into yer clog tops — ay! it's a bugger of a life, mates. Bloody black slaves was better off than we was. An' then when yer gets 'ome there's nowt for thi tay but bread-and-scrape, and thi mother tryin' 'ard not to cry 'cos thi feyther's spent all 'is week's money in a piss-up the night before. No, mates — give me soldiering every time. Even this bloody balls-up on the Peninsula, to goin' back to that life again.'

Thus he unburdened himself to us. I knew all too well how he felt, and especially the long crawl home on foot after 10 hours 'piecing-up.' I'd had some myself, except that in my case my home life was cast in gentler mould. Sutcliffe finished his spiel with a little apologetic laugh. 'Tha sounds like tha doesn't like "piecing-up",' said Cliff Stansfield,[19] grinning.

'Like it? Like it? — tha daft young bugger,' said the angry Sutcliffe, spluttering and tongue-tied in exasperation. Then he laughed 'Ay! You could say that, mate. You could say that.'

As darkness fell, we eventually settled down to sleep on our own particular patches of the firing step, most of us having completely forgotten that only about 4 feet below near where we were sleeping, Johnny Stirk was sleeping an even deeper sleep.

Such is the sanity-saving callousness of war.

---

19 Probably Pte 8824 (203410) Clifford Stansfield. See Appendix II, Notes on individuals mentioned in the text.

CHAPTER 19

A Vexed Woman

Roars of laughter in the trench sector adjoining mine made me leave the tin of early morning tea I had just brewed. Merriment is infectious. It's magnetic, too, and I poked my nose round the corner. There was a gang of chaps laughing their heads off at a Sergeant of the adjacent 'D' Company who had just wakened up and was lying there ruefully eyeing one leg of his light khaki drill shorts, now stained as if he had been the victim of enuresis. The Sergeant was a big man, virile and strong, of powerful physique, full complexioned and robust. Excess of virility and some erotic dream had stained his light-coloured shorts in patches to a dark shade. Witticisms bounced about like tennis balls, and the Sergeant enjoying the joke as much as anyone.

'Tha needs a good woman, Sarge.'

'That's wot comes through not drinking thi lime juice, Sarge.'

'Nay, it's not lime juice 'e wants, it's a good woman.'

Lime juice, issued to us daily in lieu of the absent fruit and vegetables to keep scurvy at bay, and doing the job quite successfully, too. One of the few good and necessary things we had as a daily issue, gratefully drunk by the younger soldiers, but viewed with dark suspicion by the older men — viewed as a drink sinisterly designed to 'keep Nature down', a drink to be avoided at all costs. 'Drink that bloody stuff, mates, and y'll never get a cock-stand no more, that you won't,' said our Platoon Sergeant. All the same, we drank it — and with no noticeable deleterious effects when we renewed our acquaintance once more with the bints in Cairo.

'Tha looks like tha's pissed thisel, Sarge.'

CHAPTER 19

'Nya, Ah wouldn't a' minded if 'Ad's pissed myself, but when Ah thinks about it, all this lovin' gone to waste.' He flung his head back and roared with laughter.

'Oo wur she, Sarge, thi Missus or one o' them Gippy bints?'

'Well, Ah'll tell thee one thing — it wurn't the Missus. Eee, but if she could see me now, all that lovin' gone to waste, the Missus 'ud be proper vexed. That she would — she'd be proper vexed.'

He got off the firing step where he'd been sleeping for the night and divested himself of his shorts and laid them on the parapet. Within minutes, the wet stain disappeared under the hot sun, and he resumed his shorts, contenting himself with a final laugh and a rueful, 'Ee — our Emily 'ud be proper vexed.'

Just before sundown that same day we helped him back to the same erotic couch on the firing step — in almost exactly the same place, in fact. Tenderly and slowly we eased him down on his back. His light khaki drill uniform had a similar fast spreading stain, but this time it was on his tunic, result of a stray shrapnel ball from a shell bursting over our trench. His robust complexion was drained of colour and his cheeks were tense and drawn. His eyes had a withdrawn look but apparently he was in no pain.

Heavily cheerful, one of our chaps consoled him. 'Tha'll be all right, Sarge, once we gets thee on that hospital ship,' but with the omniscience of one about to embark on another sort of vessel, the Sergeant said, 'Nay, lads. Ah'm finished.' Quietly, resignedly, he repeated, 'Nay, but Ah know Ah'm finished.'

In vain we tried to comfort him, but chaps lying on their backs in a front-line trench and with a fast flooding stain darkening their khaki drill tunics are granted special powers of perception denied to ordinary mortals like us. The view of the Styx grew clearer and clearer. Already he could hear the noise of the row-locks as Charon approached to ferry him over the dark waters.

True to the pattern of his kind, he mocks portentous Death, spitting in the face of the grisly Reaper in grim humour. His lips played in a faint smile. In a voice no more than a whisper, we heard, 'Ee, but this time our Emily reely *will* be vexed.'

Then he left us for good, to embark for... wherever it is good Lancashire Sergeants go to when their tour of earthly duty is expired.

CHAPTER 20

Money for Jam

It's a relief sometimes to turn away from some of the grimmer aspects of war, and look at some of its lighter moments. In approved Sherlock Holmes vein, let's call this one 'The Case of the Wrong Box of Jam.'

I get a lot of fun reconstructing the crime as it might have happened and although these boxes of jam were an unwelcome reality, the real history of how they came to be must forever remain the dark secret of some malodorous War Profiteer (may Allah boil him in his own jam vat!).

It's surprising, too, how little things rankle, out of all proportion to their worth. The City financier in his journey-up-town is just as much upset because his wife had burned the breakfast toast as ever he is at the prospect of the imminent collapse of his financial empire. Somewhere in the universe there must be a mysterious Law of Compensation that cuts us all down to size, to the size of what we all really are — illogical little gnomes of peevish genealogy.

Fiction demands the postulating of some bluff, hard-headed, unscrupulous North Country 'Purveyor of Jams to HM Forces,' all bounce good fellowship and double brandies. And doing very nicely, thank you. A sound type of man, this orthodox, regular Chapel-goer (believing firmly in God and twenty percent).

'Started from scratch, Ah did an' all. Self-made, that's me.' Well, at least he absolved God from the catastrophe.

'And if them lads out there, fighting for thee and me want jam, then by God, they'll have it. Even if Ah've got to put every man-jack on overtime,' and consoles himself that the profit margin will well cover the time-and-a-quarter the workmen are unreasonably demanding. 'Do some of these chaps good to be in the trenches, that it would.' Righteous indignation oozes from him like the sweat of a high fever. Another contract like that last one and he'd buy that place of Lord ——. He'd heard that the noble Lord had fallen into financial difficulties due to being away at the war and no one to look after the estate properly. 'Ah! well, war's war, and we mun all suffer. Gradely place, too, that — a real ancestral 'ome. Our Maggie's dead set 'er heart on it, she 'as an' all.'

'Beg pardon, Sir,' says the foreman, shambling in, 'but this last lot o' jam, sir.'

'Well, wot of it, 'Orrocks? Owt wrong wi' it?'

'Nay, the jam's all right, Sir, so's the tins, but we ain't got enough jam to fill all the tins, not be a few ton we ain't, Sir. And we're clean out of 'plum and apple' labels, too, Sir.'

'There's a hundred ton o' turnips wot Ah got cheap from a wholesaler last week. Wanted the money bad, 'e did. In war, we mun all 'elp one another, so Ah took 'em off 'is 'ands, cheap like. All got to 'elp one another in wartime, tha knows, 'Orrocks.'

'Turnips did you say, Sir?'

'Ay, 'Orrocks, turnips. Do Ah 'ave to spell it out for thee. T-E-R-N-I-P-S, turnips. Takes all flavours an' gives none. Nay, Ah don't know. Did thi mother never teach thee nowt, 'Orrocks? They'll pulp up grand wi' the main boiling.'

'Well, if you say so, Sir. But turnips...? And what about the 'plum and apple' labels, Sir? We're clean out of 'em.'

'Are we now? But wot about all them labels I saw at stocktaking?'

'Oh no, Sir. They wouldn't do at all. They are all labels we had for the peace-time jams — 'Blackcurrant', 'Gooseberry', 'Strawberry', etc. All these labels left over from peacetime. But no 'Plum and Apple' labels left, Sir, not a one.'

'Well?'

'Well what, Sir?'

'Dammit, 'Orrocks, do Ah 'ave to tell thee wot to do? Or does ta want me to do the job meself?'

'Well, if you say so, Sir.'

'Ah do say so. Now be off wi' thee and get thi job done. They're sealed cans, aren't they? And if owt's ever said, anyone can make an honest mistake, can't they? And dammit, — 'aven't Ah got a lad o' mi own, fighting in the trenches?'

This lad of his, a God-given sop to his conscience. Not a bad fellow, really. A good officer, fair, if a bit strict, but dead fair. But you couldn't warm to him like you could to some of the other officers, even if they weren't so fair and straightforward. Some of these officers would lose their tempers and fly off at the deep end — just like we do sometimes, and then ladle out the punishment a bit steep. But somehow you liked the buggers in spite of their faults. They're sort of 'human', if you know what I mean — got faults, weaknesses, just like you and me. Guess that's what makes us like 'em, the bastards! But this other bloke — coo! Like I said, dead fair and straightforward, but a bit 'toffee-nosed.'

Well, I dunno! There's always 'summat' that makes a man a bit different, just like there's always summat that makes one elephant a rogue. I should think that what made this chap so — ah! — what's the word I want — so 'aspiring,' so 'snooty,' was the handicap he had been born with. And after all, 'Sidebottom' — even if it is a good old Lancashire name, it's a hell of a name if you intend to put in for a commission in HM Forces.

'An' wot's wrong wi' Sidebottom?' said his masterful father. 'What's wrong wi' Sidebottom, Ah'd like to know. It's Sidebottom money that gives thee tha good 'ome and 'as sent thee to that posh school. It's Sidebottom money wot's learned thee to talk proper. So wot's wrong wi' Sidebottom, lad? Tell me that.'

'Oh! There's nothing wrong at all, Father, with Sidebottom. It's good old Lancashire name, I'll grant you. If ever you buy that place of Lord... you know, Dad' — (cunning young fellow this one, knows how to probe the vulnerable spots. Should go far, with or without that name).

'Still, we can't do nowt about it, can we lad? We're born wi' Sidebottom and we're stuck wi' it.'

'Not necessarily, Father.'

Then unfolds this young officer-to-be his plan.

'It doesn't cost much, Dad — one or two letters altered, y'know. The name's the same, merely pronounced a little differently you know — and with the accent on the last syllable of the name.'

'Well, Ah don't know nowt about syllables and all that stuff, but if y'thinks it'll be all right, lad...'

So 'Sidebottom' becomes 'Siddey-botham.'

'Sounds dead posh, don't it Maggie?' says the benevolent manufacturer o' jams and patriotic contract-wangler. 'Eee, but our lad's got a 'ead on 'im all right. E's a true Sidebottom — er — a true Siddey-Botham, ain't he, Maggie? And if Ah gets that little place of Lord ——'s, and 'appen someday a knighthood, like, it'll sound all right, won't it, Maggie? Sir Joseph Siddey-Botham, Esq., Bart., "for services rendered during the Great War." Not that Ah'm denying Ah've made a bob or two out of it. Ee! but it's bin 'money for jam' as the saying goes — this war. Eee, but that wur a good 'un, wurn't it, Maggie, "money for jam",' and laughs heartily at the double entendre.

We actually did have for a short time in our trenches a Lieutenant Siddey-Botham (née Sidebottom). Cad that I am, I use unfairly his name as my inspiration for an excursion into fiction to explain our own particular jam debacle. Whether this officer came of good old British jam-making stock, I haven't the faintest idea. (If I malign him thus, I humbly beg his pardon.) True, he was a bit upstage, 'thinks his shit doesn't stink,' said one of our coarser types, but with a name like that, what would you? And all of us go through a bit of a peacock phase in the course of growing up.

But the patriotic jam-maker himself, well — I warned you earlier I have a shocking imagination.

And it was no imagination when the actual tins made their appearance on the Peninsula. Whether we like it or not, we now have to leave the beguiling land of fiction for the cold tundra wastelands of hard facts.

The bi-weekly case of jam our Sergeant was opening aroused no enthusiasm in us — except for the piece of wood the box was made of — handy for lighting a little fire for brewing a can of tea, this wood — nice dry kindling wood. The known contents of the box was such that we didn't dwell on thinking about it too long — these tins of green frog spawn masquerading under the name of 'Plum and Apple'

Jam. We had grown to loathe it, and God knows — we of all people weren't fussy eaters.

It was the Sergeant's long whistle of surprise that first raised a gleam in our dull and unenthusiastic eyes. 'HELLO, HEL-LO, HEL-LO,' he says, loud and slow. 'Wot the 'ell we got 'ere. Reckon they've give us Orficers jams by mistake. Best take it back, save a bloody row, I s'pose.'

By now we had crowded round the opened case and glimpsed the Treasures of the Incas. Neat little rows of tins of jam of all sorts — 'Blackcurrant', 'Raspberry', 'Strawberry', 'Gooseberry' — anything you liked. You name it, we had it.

'Nay, you can't take it back now, Sarge, wot wi' the box bein' all broke.'

'Nay, Sarge. They made the mistake — they mun stand by it.'

'Nay, Sarge — 'bout bloody time we 'ad summat decent to spread on these f— biscuits. Nay, you can't send it back.'

The chorus of protests mounted. Still our Sergeant hesitated. 'A right regimental bugger, our Sergeant,' one of our blokes muttered despairingly. 'Ah'll bet you he sends it back.'

But the Sergeant for once broke his code — either broke it or adjusted his military conscience — like the RCs do sometimes when they want to do 'summat' the Church forbids and yet keep their righteous self-respect. Only the military conscience is of greater flexibility, and follow the scientific law of Infinite Variability as governed by the Rule of Alternating Stresses and Strains. You'll probably find a text-book on this subject on the science shelf of your library at home, if you look! So our Sergeant short-circuited the problem and QED-ed it by saying, 'Once we've ate the bugger, they can't do much about it. So get it inside thee, lads, quick-like. 'Ere, Corporal, wot's thy fancy — blackcurrant? 'Ere y'are then, coming over,' and he's busy lobbing over tins of jam according to his grateful client's needs.

'Wot's thine, Cobber?' he calls to me, 'Blackberry — ee, but Ah thinks tha's unlucky. Nay, howd on a minute, 'ere y'are, Blackberry coming up.'

All men have a secret vice — or weakness, if you like. Mine's Blackberry Jam. For a tin or jar of this delectable conserve, I'd sell my own grandmother down the river and my children into slavery.

'God is good.' 'Blessed be Allah.' I thanked 'em both, just to be sure.

It was Jimmy N—— who first cracked open the secret. He'd already dug his bayonet into the soft tin top of the can and hooked out a bladeful. But as I told you before, nothing ever bothered him much, just guffawed, 'Reckon some silly little bitch at the factory put the wrong label on this bloody tin. It's 'Plum and Apple', not 'Strawberry' like it says on the tin.'

Other tins began to be opened and there was a dead silence. I didn't dare open mine. I'm what they call a moral coward. Even if I have a bet on a horse, I don't dare look at the results in the Stop Press, and have to wait until the bookie comes round sometimes and begs me to take the winnings!

The silence was a bit of the silence of sudden shock — a shock that there could exist anyone in the world who'd play such a trick on blokes like us in our own poor straits. A few mutterings later, then despair, hate, disillusionment, cynicism brittle as glass — all the things that men are better off by being strangers to — they all crystallised in black devil of anger. But most of all there seemed to be a sadness, a sadness that anyone would play such a trick on soldiers at the front. As our Sergeant said, and more in sorrow than in anger, 'We may be a dog-gone, down-and-out, scruffy, lousy lot of bloody bastards of front-line soldiers, but we didn't deserve a trick like this. Ah reckon all jam-makers is bloody bastards.'

The passion for British justice aroused one of our corporals to defend the jam gentry. 'Nay, Sarge, there's good and bad, tha knows, everywhere. Ah reckon there's plenty of good jam makers somewhere.' The Sergeant agreed, adding bitterly, 'Ay, there is... in the cemetery.' Then we flung the tins of this revolting pseudo jam over to the Turkish lines like hand grenades. I often wonder if the Turks ever got 'em.

We had all contributed our share of individual condemnation, and our good wishes for the jam manufacturer in his future life — all except our little Irish boy. 'Tha's not said nowt thiself, Paddy — wot's ta think about it?' Paddy was fingering his beads and seemed absorbed in some problem of the Infinite, and was recalled to consciousness by the question. Lacking the raw raucousness of our own harsh dialect, his speech and tone of voice was habitually soft — deceptively so when most angry. As he purred out his own particular anathema, it was in a soft Irish voice of caressing quality more appropriate to a lustful lad of Connemara trying to persuade the reluctant colleen to take the short cut home through the woods because of the danger of falling into the bog now that it's dark. If you couldn't catch the words he was saying, you'd have never known that by comparison he was making the curse of Cromwell sound like the lisp of a baby's prayer.

Frequently calling on the deity, the Virgin Mary and the Blessed St Patrick, he poured out his heart fulsomely. 'Holy Mother of God,' he said, invoking our Blessed Lady of the Immaculate Conception, and crossing himself piously, 'if I could lay my hands on that…' (regretfully there's still a law on obscenity! ex-servicemen please fill in the blanks) '…but if I could lay my hands on the spalpeen who did this to us, I'd…' (and then followed a list of anticipatory sadistic tortures that did credit to Paddy's imagination and inventive powers).

An Irishman with a grievance is never single-minded, and doesn't stop at the one. He embraces in his denunciation all the ills and troubles that from the beginning of time have bothered his shamrock-ridden land.

'God forgive me for what I say, but 'tis only a spawn of an English spalpeen who could play a trick like this on a bhoy from the Ould Sod — god forgive me for my words, and you being such a decent lot of bhoys, even if you are naught but a lot of rotten Protestant English, bad cess to ye.' Then he went on to list a thousand-and-one grievances that every Irishman treasures up like precious pearls — (Begorrah! but he wouldn't part with even one of them, no — not for all the tea in China), and ended up by tracing the ancestry of the Brotherhood of the Boiling Jam Vats, and in particular of the japer who had played this cruel trick on us. Descendants of a long line of bastards, suckled by poxy whores and fathered by randy 'commaircial' travellers, he went right back to the Adam and Eve of the species who, he said, were a syphilitic she-dog and a 'clapped-up' mountain goat. He then wound up in a wheedling tone, 'And I don't suppose none of ye bhoys, daysent lot of lads that ye are an' all, would have a butt of a cigarette ye're not after wanting now, would ye?'

His grinning audience shook their heads. He'd have had our last butt if we'd had one. It was worth it. One of our lads said, 'Nay, Paddy, but if Ah 'ad any Ah'd give thee a whole packet — after that lot, that Ah would.'

''Tis a daysent lad ye are, me bhoy, even if ye are naught but an English spalpeen, God help you. Like me own brother ye are, except for being English — and that ye can't help, God forgive you.'

Every body of men should include an Irishman in their midst. They say they make the best engineers, the best roadmakers, the best bridge builders in the world. I don't believe a word of it. But they certainly make the best human beings and preserve the sane English from the madness of sanity.

In the Catholic Church of which I am such an unworthy and backsliding member, we have a lovely little prayer. A real genuine one, said aloud in Church in memory of someone of the congregation recently 'passed over.' It is always said by everyone with genuine sincerity, one reason being the normal decency of folks and the other, that they know it'll be their turn next for the same prayer to be said for them. Even scallywags like me, threatened with the dreadful retribution of boiling brimstone if I don't mend my ways, even I intone it reverently, in the hope that the vats of brimstone won't splash on me too much when I eventually pay a visit to the Boiling Oil and Brimstone Factory awaiting us.

'May the souls of all the faithful departed, through the mercy of God, rest in peace.'

It's a nice prayer. But I hope the most devout Catholic soul won't object if I alter it ever so slightly, just for the benefit of that Most Unholy Band of War Profiteers. I call it the 'Plum and Apple' prayer. 'May the soles of all such bastards as these blokes, tread the red-hot grid-irons of Hell.'

CHAPTER 21

My Lady Nicotine

About a week after this jam debacle, Mafeking was relieved. History was made. The Heavens opened and manna showered down on us like hailstones. We got our first issue of cigarettes.

There was such a bloody noise from the cheering soldiery that the very gods themselves could be seen peeping down from behind the clouds to see what-for all this blasted commotion. Snooty and superior from their lofty vantage point, they derided our childish human raptures. But they'd never known themselves that 'creep-up-the-wall' feeling for a fag, the yearning to suck deep down again into our lungs the soothing smoke of the drug.

Good fags they were, too, 50 a week per man — a new brand to us, Ruby Queen. Non-smokers amongst us suddenly attained a popularity they'd never envisaged!

Napoleon might have been right when he said an Army marches on its stomach. But it fights a lot better if sponsored by the Imperial Tobacco Company!

Our reverent Irish boy had been following his morning custom of propitiating the gods by telling his beads, when the fags arrived. He interrupted one of this three 'Hail Marys' to ask for a light. 'Hail Mary full of grace, the Lord is with thee (An' will ye be giving me a light for me fag now). Blessed art thou among women (puff, puff) — (begob — but it tastes good), and blessed be the fruit of thy womb, Jesus.'

Yet I feel sure our Blessed Lady will have pardoned him just this once. Ten weeks without a fag is a long, long time. He crossed himself reverently. 'God is good,' he said devoutly, and his irreverent companion grinned. 'Ay,' he said, 'but a fag's better.'

Rumours of an impending fag issue had been circulating for some days beforehand. Me — I reckon the authorities do this sort of thing deliberately, circulate the rumour first and then accomplish the event. They do it this way, maybe out of consideration for our weak hearts.

It's quite respectable having your name on a War Memorial 'Killed in Action' or 'Died of Wounds,' but it wouldn't look so good, 'Died of shock at the first fag issue.' But even when the fags arrived we sniffed suspiciously. The memory of the last jam issue still rankled.

'All Platoon Sergeants to report to CQMS Stores.' When we heard this, twittering hope fluttered our tiny hearts. Maybe fags at last. Quien sabe? And eventually our Platoon Sergeant getting us conveniently around him and clowning like a grotesque Santa Claus, holding a large packet of 'summat' behind his back and grinning like an ape. In words of noble Greek oratory he addressed us.

'You crummy lot of louse-bound bastards, the authorities in their goodness of 'eart 'ave seen fit to issue from now on 50 fags a week per man. Now line up you so-and-so's for a tin of 50 each, and the first man wot tries to work the double-shuffle on me and lines up twice, Ah'll brain 'im with me own entrenching tool handle.'

So the contented soldiery received their first issue of cigarettes in nice airtight tins of 50. We hadn't got a match or a light between us, but one of our chaps said he'd fix it with a piece of puggaree. A puggaree being that length of chiffon that drapes itself so artistically around the pith sun helmet (as if you didn't know that already, you hard-baked 'old sweats'). He rolls a piece of puggaree into a tight string and sprints to the cookhouse a quarter of a mile away and where there was a fire. This puggaree stuff, if it's rolled tight and lighted can be kept smouldering like a piece of smouldering string, and keeps alight for a long time if one keeps on blowing it. Thus the Olympic torch was brought successfully to the waiting soldiery craning their necks for the arrival of the runner and m no time at all the incense of tobacco spiced the morning air.

Our Platoon Sergeant was nothing if not dead straight and fair. There were seven tins surplus to our numbers — due to seven blokes having been bumped off in the meantime since the last strength return had been rendered three days ago. This over-issue of seven tins he discussed with us, whether to share them out equally, or to save them to give a tin to the next poor sod wot gets wounded. "E might find 'em handy, like. There's nowt like a fag for comfort when you've stopped a bullet.'

There was no card vote for the answer, no show of hands, and no secret ballot as to the proper disposal of these precious extra fags. Just a unanimous growl of decision — 'Ay! Keep 'em for the next poor bastards as'll need 'em more'n us.'

That's what I like about my crummy, lice-ridden mates. They might try to steal some of yours if they were short, but there wasn't a man who wouldn't have cut off his right hand sooner than touch any one of these seven tins. The conscience of a soldier! Someday, mebbe, and when I get round to it, I promise to draw you a diagram of it!

From this first trickle of an issue the supply of cigarettes became a flood. Hard-working, good-hearted people back home had started a 'Comforts Fund' for the soldiers in the trenches and soon we had more cigarettes than we could cope with — right to the end of the campaign. I often wonder if enough credit's ever been given to those blessed people at home who organised this regular supply of life-giving fags. It's a bit late, I'll agree, to start saying 'thanks' but somehow during life there never seems to be enough time to get round to saying 'thanks' to all the people who've helped you — doctors, nurses, parents, relations, teachers, friends, senders of fags to distressed warriors — oh! there's no end of people we're all indebted to, everyone of us. That's one of the reasons why I think there's just got to be an afterlife, if only so as we can meet up again with these folks and say, 'Thanks.'

The pipe-smokers, they fared badly. The choice of the weekly tobacco issue was either 50 cigarettes, or two ounces of pipe tobacco — again in nice little sealed airtight tins. To a pipe smoker, there's no substitute for pipe tobacco. All the free cigarettes in the world are as naught to him. Cigarette smokers are irascible 'junkies,' their frayed nerves only soothed by sucking one of their white sticks. But a pipe-smoker: to him, to smoke a pipe is almost to take part in a religious rite, an act of profound devotion. To see their faces suffused with the calm joy of a well-drawing pipe is to gaze upon the face of a saint caught up in ecstasy. Calm, peaceful types, these pipe smokers, contemplative, philosophical, dignified and proud. Denied his little white sticks of comfort the fag smoker is jumpy and irritable. Shocking types, these fag smokers, the real 'junkies' of modern times. I should know! If all the fags I've smoked were put end-to-end, well — they might not quite reach to the moon, but oh boy! — wouldn't it make a lovely long smoke! But the pipe smokers, denied their weed, they don't get irritable and touchy like us, merely look a little sad and come over all withdrawn and distant.

It was great to see them looking happy again, almost human again they seemed. Some even regained the power of speech. 'Ruby Glen,' said one of them, reading the label on the tin. 'Never smoked that brand before, but as long as it's 'bacca…'

Ay! As long as it's tobacco. Then they opened their two-ounce sealed tins and explored the green soggy leaves neatly pressed together within. One even cut some up and tried to smoke it and we were able to add yet another collection to our already rich vocabulary of cuss words.

It was a lousy trick. Tempting no doubt to the manufacturer or whoever it was who took advantage of the sealed airtight tins to pass off this sodden mess of leaves as real tobacco. But as the Good Book sayeth 'What shall it profit a man if he gain the whole world, if he also earns the heartfelt curses of decent men like pipe-smokers.'

To these gentlement of the quick buck and the sealed airtight tin trick, the 'Plum and Apple' prayer — 'May the soles of all such bastards, etc.'

Fortunately, about a couple of days later it gave me a chance to pose as the world's greatest philanthropist! Quite a mean little skunk I am really, but to the pipe-smokers of our Platoon I was the Golden Boy direct from Heaven. 'You're the nicest lad in the whole bloody platoon, Cobber, buggered if you ain't,' one tells me. 'The sun itself shines out of your arsehole, that it does,' says another. I got so many verbal testimonials as to my character, the halo got so tight I got a headache and was the start of me developing later on in life into the world's most conceited bastard. And all this due to the fact that my father, then in Canada, had sent me a couple of pounds of really good Canadian Plug Tobacco. Apparently in Canada during the first war, tobacco was duty-free *and* post-free, if addressed to a member of the Forces at the battlefront, so I got a bumper parcel. Inside was a typical little 'Dad' note — 'If this is too much for you to smoke in one night, you know what to do, lad. And keep your head down.'

So this phoney Lord of this scrubby little manor goes round, scattering largesse to the wretched peasantry of pipe-smokers. One of the grateful recipients was a rough, tough ex-navvy from Civvy Street, a quiet sort of bloke generally, about 40, and with a big black walrus moustache. King of all the pipe-smokers, this chap. 'And ain't yer smokin' some yourself, lad? I know yer a pipe-smoker sometimes.' Sadly I explained to the ex-navvy that I'd broken my one and only pipe and hadn't a hope in hell of getting another in this bloody place, whereupon he delved to the bottom of his pack valise and produced one. 'I was keeping it for me own lad, he said. "Appen 'e won't be needing it now.'

It was a beautiful pipe, well-seasoned and with a large briar bowl and curved stem — a real Sherlock Holmes pipe. It was as familiar to the boys of our Platoon as our own faces.

Back in Cairo in our training days and when we used to relax in the evening sun on the barrack square of the Citadel Barracks, watching our Drum and Bugle Band 'Beat Retreat' every evening, this chap would be sat with us. Squatting on our haunches up against the wall, we'd watch, and never fail to delight in this evening display of military pageantry, watch the bugles and drums make their brave and glittering display, kettledrums rising and falling rhythmatically from the leopard-skin covered legs of the drummers as they marched to and fro, weaving intricate formation patterns. Ah! they'd make a brave show as the bugles and the drums caught the rays of the setting sun. I'm a real nut for pageantry, especially military pageantry.

And we'd watch intently. But not this chap — absorbed as ever he sat with us, in carving or etching on the wooden bowls of his pipe some little thing that took his fancy. Once it was a set of minute cherubs, exquisitely carved and beautifully finished. One of our Officers offered him any price he liked for it, so characteristically he gave it to him and resolutely refused to take a penny. Another pipe sported a hunting scene and the one he gave me was a colourful etching (or whatever you call this sort of carving) of the Sphinx and Pyramids. In the middle distance an Arab Dhow glided peacefully by. A fine piece of work by any standards. 'This is for the lad when Ah gets 'ome. 'Appen, too, it'll remind me of this place.'

CHAPTER 21

'You and your bloody pipes,' said one of the lads one evening, exasperated that Tasker had turned a blind eye to a particularly good display of the Drums and Bugles.

'Thee and thi bloody drums,' said Tasker imperturbably, and went on with his carving.

So I accepted the pipe under protest. I have a native genius for blundering into the minefields of life. 'But what about your lad. I thought you were saving it for him?'

'Nay — e'll not need it now.'

'You mean … ?'

'Ay, lad. Ah mean just that.' And he stared blankly at the trench wall. After a bit, 'When did it happen?'

'Last month — about five weeks after we'd landed 'ere. Torpedoed.'

He was silent for a bit. Then his eyes lit up proudly. "E's in the Navy. 'E's an ERA Petty Officer, too. Looks damn smart in 'is uniform, 'e does.'

His present tense intrigues me. The penny drops. It drops damned slow for me sometimes. 'Ah!' I said, much relieved, 'wounded, you mean.' He stares at the trench wall again. 'No,' he said. 'Like Ah told thee: torpedoed.'

The present tense has obviously grievously misled me. I'd forgotten to allow for the Eternity Continuity Factor in a father-son relationship. There's no past tense with a fond father. It's a 'forever' business with him. Tasker's lad is torpedoed, his son's gone on a long, long journey. But to Tasker his son is still here.

It makes sense.

He puffs furiously on his pipe, then slides his head sideways, away from me, and seems to get a bad fit of coughing. 'Ee, but this bloody bacca's strong, Charlie,' and stems a cruel gush of tears with the inside lining of his soft-cloth service cap.

'Yes. I agree. It is a bit strong.'

'Fair made mi eyes watter, this 'bacca. Ee, but it's strong, it is an' all.'

'Yes it is, isn't it? Rather strong.'

'Good 'bacca, mind you. But strong, a bit strong, doesn't tha think?'

'Ay! Canadian tobacco. Always a bit strong, I think.'

'Ay! good 'bacca, Charlie. But a bit strong, like. Fair makes yer eyes watter, it does, if yer swallows a puff the wrong way.'

(God damn the cult of the stiff upper lip. Any bloody fool can laugh. It takes a good man to cry.)

Later, the atmosphere a bit easier, I asked, 'Do any of the other lads know about... y'know?'

'Nay, Ah've not told nobody but thee, and that wur accidental, like. As to t'other lads, doesn't tha think they got enough troubles o' their own, on this bloody place?'

So this rag-a-muffin Lord of the Manor, this pinchbeck Prince Bountiful leaves his peasantry, and his stinking largesse. 'Proper humbled' he creeps away to his own trench corner.

CHAPTER 22

The Cobbers

'A peninsula is a narrow neck of land almost entirely surrounded by water.' So runs the school text book.

For me, a peninsula will always be a two-bit corner of the Eastern Mediterranean, peopled by mongrel troops of the British Empire. Like most mongrels they tended to retain strong traces of the best characteristics of their mixed pedigrees. The more numerous British, well, who in the world could begin to sort out their own particular Cruft's qualification for this gigantic dog-show. A friendly, tail-wagging bunch of tykes, these — anxious more than anything in the world to please, and to be pleased, but ready to fight to the last mongrel cur if necessary. Their lingoes and dialects were as mixed and varied as was their own colourful historic lineage.

The Scots — a great breed these — tending to greater purity of pedigree. Good reliable types if you got on the right side of them, but maybe a bit touchy — and with a tendency to snap if you upset them.

The Irish, another purer breed, a bit unpredictable, loving a romp and a dog-fight just for the hell of it, even amongst themselves.

The Welsh? Well — damned if I could ever classify these chaps. You've got to be a Welshman to understand a Welshman. Insistent they are, too, in their claim to be the original and first-ever British stock, long before the English even had an identity. What's that favourite saying of theirs — 'Wales was Wales before England was was.'

Each with their own prides and traditions, these mongrel packs — prides and traditions cherished lovingly and oft vaingloriously.

Take my own gang of Lancastrians. For us, there are of course, only two sorts of people in the world — the Lancashire lads and the other poor fish — 'the others' — those who didn't have the luck to be Lancashire bred. Oddly enough, the Yorkshireman thinks erroneously along the same lines.

In my own platoon we had two sections of Lancashire lads and two sections of Yorkshire chaps from just over the border, and daily we carried on where the Wars of the Roses left off.

'You crazy lot of coots,' said our exasperated Company Commander, one day. The ringleaders had been hauled up in front of him for unseemly behaviour in one of the main streets of Cairo. I forget how it all started — this particular bit of Red and White Rose rivalry.

'You crazy lot of coots — if the British Army spent half as much energy fighting the enemy as you lot do in fighting amongst yourselves, the war would be over in a week.'

His twinkling eye belied his stern face.

'Get 'em scrubbing the Canteen Floor, Sergeant Major — half each to the Lancashire and Yorkshire lot.'

'All right, dismiss.'

I never did see a canteen floor so clean after we'd finished the job. We were still arguing, long after Lights Out as to which was the cleaner half. But on the battlefield we clean forgot the Wars of the Roses and concentrated on the job in hand.

Ay! It's a great military idea to have a composite and mixed force of the different breeds of British stock. Rivalry — the 'one-upmanship' element — to do better than the other lot — it makes good fighting material; what a pity the campaign wasn't better organised. With blokes like these, we could have gone right through to Constantinople like a hot knife cutting through butter.

An off-shoot of these mongrel packs but now so far removed geographically and historically from the parent packs as to be a distinct breed of their own, were the 'Cobbers' — the eye-catching mastiffs of this gigantic Crufts Show. I don't think I ever met a better bunch of boys I liked better on Gallipoli than the Australians and New Zealanders, and I don't suppose I ever shall — anyway, not this side of the River Jordan. Massive, colourful types, this first Australian and New Zealand contingent who were with us on Gallipoli. Good natured, easy-going, and with the mastiff's unshakeable bravery and steadfastness. But what I liked about 'em most was that either by some accident or breeding, or the life they had been leading, they seemed to be purged of all human 'littleness.' And this element of peevish human 'littleness' tends to be a bit strong sometimes with us who are more closely bred. There was, too, an aura about them of an honest and robust earthiness that I found refreshing. No angels, these boys. Discipline was a dirty word to these wild and rugged individualists. If they had a fault as soldiers, it was that they had never been taught, or were incapable of understanding, that a needlessly dead Australian soldier is no good to a fighting force and many, many a good Aussie bit the dust needlessly through his own individual desire for prowess and not to be bested by the enemy. This desire for individual prowess was perhaps inseparable from the very pattern of their lives back home. How else could they have carved out their empire on the other side of the world?

These blokes really fascinated us — not so much their colourful feathered hats, not so much their powerful physique or their queer twang of speech, but most I think by their lordly bearing and a lofty disregard of danger and death. Trench life they found irksome — too cramping for their excess vitality and their desire for activity — even at the cannon's mouth. A vigorous spell of leap-frog in the open — even at the cost of a few needless casualties — well, it was worth it. 'Can't expect Austry-lian boys like us to stay cooped up in a trench all day, can yer, Tommy?'

Most of them were a little older than us, between 25 to 35 seemed to be their average ages, as against our 19 to 22 or so. More matured, they were more rock-like in adversity and their native strong constitutions weathered the vagaries of material hardship much better.

There was only one day in the calendar for these Aussies and that was today. Today to be lived. Maybe there's no tomorrow for a lot of these Australian boys, so let 'em live for today. No yesterdays for them either, except perhaps when they'd audibly yearn 'just to see them tall gum trees again back in Austry-lia, and spend a few days in the Bush. Ever seen them gum trees, Tommy? They're that high, you'd think they'd bloody near touch the sky. Lookin' at 'em, Tommy, makes yer feel sorta, if you follow me — good. And a few days in the Bush, Tommy, makes a man feel free, it does. Y'can't beat it.'

The passionate yearning in his voice as he gazes into high space moves you strangely — paints a vivid picture of the 'fullness of nothingness' of the Bush. Like you get when you're in the desert — surrounded on all sides by nothing. And the nothingness is more real and tangible than is matter itself.

Maybe I don't describe it the way I should, but those of you who've lived a few months in the desert, you'll know exactly what I mean. And know, too, how a man feels. Me — I sure knew what that Aussie was yearning for. It could 'a been me, but the only tall gum trees for me, back home, were the forests of tall factory chimneys.

'Lookin' at 'em, Aussie, makes yer feel, sorta, makes yer feel, makes yer feel, feel like bloody 'ell, if you follow me.'

And when your dawn silence of the Bush is broken by the screech of cockatoo, or whatever sort of damned birds you got there, my dawn was broken by the sirens of those tall chimneys calling me in for 10 hours slavery every day, watching those millions of imbecile cotton bobbins spinning away for dear life.

'Makes a man feel free, it does. At least, it does come Saturday mid-day and when you've finished your slavery for the week, Cobber. Them's my yesterdays, Cobber. No tall gum trees.' But it doesn't stop me from feeling what you mean, my little Aussie.

Back in Cairo in our training days, the Aussies were under canvas a few miles away at Mena Camp, in the shadow of the Pyramids. They spent their abundant and surplus energy in wild and oft drunken orgies in Cairo some nights. Officers and Privates would be seen by shocked British soldiers hobnobbing together, discipline and military etiquette being quite unknown to them. 'Saluting's a thing for Pommie bastards, not for Austry-lian boys like us.' But still, we had a great affection for the 'Cobbers', these wild Australian boys. The only slight resentment we ever felt — and even that was not their fault — was the fact that their basic rate of pay was 6/- a day as against our 1/- a day. But there it is — the wages of war, like the wages of sin, quite often vary!

The indiscipline of these chaps was not so much deliberate as a complete ignorance of King's Regulations and Army Council instructions, and the Manual of Military Law — those two formidable books that have governed the way of life maybe ever since Julius Caesar's time, for the British Army. It was an ignorance so abysmal that British

Orderly Rooms Sergeants from the Regular Army and well versed in the niceties of the application of discipline, were sometimes attached, one to each Australian battalion, to advise and instruct the young Aussie Officers in the British Army's traditional way of life. And thereby hangs a tale. There were many stories about these Aussies that we treasured gleefully — some wildly impossible, some I could tell you over a quiet pint and with the barmaid well out of earshot. Quite a lot of these stories were invented for, to us the Aussies were the fabulous ones — the ones about whom it was natural we should weave stories. There's one story I particularly like. If it's not true, then it ought to be. (No, Auntie, don't skip a few pages. It's quite clean — well, nearly!)

The GOC of these Aussies, sick and tired of complaints about his troops' disorderly habits, put out an all-party blast right along the line that it had to be stopped and proper punishments meted out in future to all offenders against military discipline, etc.

Punishments for military offences vary. Often they remind me of that little ditty of Gilbert & Sullivan's 'Mikardo' —

> The object all sublime, we shall achieve in time,
> And make the punishment fit the crime,
> The punishment fit the crime... etc.

But they're reasonably fair, these punishments — tending if anything to leniency, and obviously thought up by a man wise to the peccadillos of thoughtless but not necessarily evil soldiery. They could range from a simple offence like 'drunkenness', for which the usual punishment was a three days stoppage of pay, plus being confined to barracks for three days. Three days confined to barracks, plus a mere fine of 3/- doesn't seem much, but 3/- deducted from a weekly wage of 7/- hurts more than you think. One of its lesser horrors is a period of enforced teetotalism. This particular punishment was loosely and inaccurately referred to by the soldiery as 'being given three days' pay.' So, if you were 'given three days' pay' it meant, actually, you'd been given the punishment of three days stoppage of pay, plus the accompanying curtailment of three days liberty after duty. For more serious offences you could get — oh! — anything up to 28 days or 56 days detention — which means being incarcerated in the Glass House, the nickname of the Military

Prison, during which time the staff there do their best to make you wish you'd never been born — a type of prison that I often think must proudly trace its ancestry back to the Spanish Inquisition. The proud boast of these prisons is that they can even tame lions there. I believe 'em, the b——s. I should know!

These serious offences were rare, but we did get 'em now and again. Every soldier serving abroad sometimes gets a touch of the 'Cafard' and goes a bit berserk. Or it could be you're your Mum writes and tells you that your wife whom you last left in England some six months ago is now three months pregnant. So what else is there to do but to go and bash some passing NCO just to relieve your feelings — and spend the next 28 days 'inside' wondering why you'd ever done such a damn silly thing!

But most offences were of the same minor and dreary pattern.

Such as when you're feeling exhilarated and on top of the world after a few pints. And some snotty-nosed Military Policeman stops you and tells you to button up your tunic and to put your hat on straight and not wear it back to front. So naturally you tell him things, such as his father must have been a bachelor, or something like that.

Mind you, soldier, most of us would agree with you — about the MP, I mean — but it's not diplomatic to voice your opinions in a public street, nor to offer to fight him for a pound, cash. So you're given three days pay as a punishment. Never mind, you've still got 4/- left to see you through the rest of the week. But you're not so lucky as the Aussie we heard this story about. And when he was marched in front of his Company Officer. This particular Officer was feeling sore. In fact, all the Officers of his battalion were feeling sore — they'd all just had a good rousting about not keeping discipline. The delinquent was charged with being drunk in the main street of Cairo the night before, and after the formality of evidence had been read, the young Aussie Officer pronounced sentence — '28 days detention.'

There was a gasp from the advisory British Orderly Room Sergeant who sat in in all these proceedings and a hurried plucking of the young officer's sleeve and a whispered urgent conversation.

'You can't do that, Sir, not for a simple "drunk" — and a first offence at that. The very most you can give him is 'three days' pay.'

Whereupon the red-faced and embarrassed young officer dug his hand into his pocket, counted out 18/- and pushed it over to the delinquent. 'There you are, you bastard — three days' pay. But I warn you — next time you appear before me you'll get fuck-all.'

Careless with their lives, these blokes, careless and generous. Bewilderingly so. The gift of Life is a thing to treasure, not to squander away. If the fortune of war robs you of it — well, it's just too bad. But every soldier has a duty to take reasonable precautions and not to stick his neck out unjustifiably and just for the hell of it.

I was on a special and lousy, thankless duty one day in a little 'sap' (or small trench) abutting forward at right angles from our front line. At the end of this little sap — about 20 yards in front of our front line — was fixed a sniper's hidey-hole — an inch-thick steel plate with an aperture in it just large enough to poke a rifle through. You swivelled aside a little piece of the same inch-thick steel, and designed just large enough to overlap the aperture when not in use, poked your rifle through, took a quick sight at anything moving in the Turkish lines ahead, and try a quick shot, making sure you're back to safety and with the aperture closed again, and all this before you can count ten — otherwise you'd get one back through the aperture. For like I told you, these Turkish sharp-shooters are terrific.

All around the aperture the steel plate was dented by Turkish bullets that had thudded up against it, and even the piece of steel that swivelled over the aperture — even that was dented and bashed. After two or three hours of this duty and if you survived your first experience of it, you knew almost to the split second how long you could remain poised for a shot and be able to dodge back safely again. It was a job nobody really liked, but if you were doing it, it became a sort of challenge and in a weird sort of way you found it rather stimulating.

But if the bloke on duty lingers overlong in taking his pot-shot at the Turks, then some Mum back in England will be taking a hurried trip to the local dressmakers — 'Mourning Orders Promptly Executed' — to get fitted with one of those shiny black dresses. And she'll listen uncomprehendingly to the Parson when he calls around that evening to comfort her, and talk about it being 'God's Will.' Bitter and twisted, she, in her simple way will wonder how can it be the will of God that all this should happen to her to lose her only son, and him being such a nice lad an' all. For grief affects us that way.

But don't take these Parsons too seriously, Ma'am. They're a right lot of ... so-and-so's. An earthquake — and thousands crushed and buried; a flood — and hundreds drowned; an air disaster and many burnt alive; a four-year old and blue-eyed darling knocked down by a bus ... It's all the 'Will of God' to these Parsons. But don't you believe 'em, Ma'am. This God of Abraham, Isaac and Jacob — some of us — and with the 'stars still in our eyes' — we still like to think of Him like you do, Ma'am — as an old-fashioned God of Love, and not as a sadistic monster obscenely planning these catastrophes.

All this rigmarole, of course, is just my long-winded way of saying that if you don't want your own Mum to get mixed up in all this damned nonsense, you've got to close that aperture a bit smartish, otherwise you'll get it where the chicken got the chopper. Like those two Aussie blokes one day when I was on this lousy bit of sniping duty. I still feel a mite responsible that two unknown Australian Mums got mixed up in all this stuff, but if you were only a Private Soldier like me, you can't win — not ever.

Two NCOs of the Australian Artillery — huge chaps — about 6′ 4″ and broad with it. They were seeking an Artillery observation post as far forward as possible for the next offensive of ours. That's how they came to be in front of our front line and chatting with me. One was a Staff Sergeant and did all the talking and the other, a Corporal, just listened and didn't say a lot.[20] The Staff Sergeant grew intrigued with my job and wanted to have a go himself but I told him the score and that if you didn't know the trick, you'd be a dead duck before you could look round. This seemed to peeve him — 'didn't want any Pommie to tell him all about shooting.' 'Back in Austry-lia he could shoot the toe-nail off a 'possum's left foot.' He was a genial smiling bloke of about 35 to 40 with a huge fair moustache, and pleasant enough to talk to, so I tried to kid him into talking about life 'down-under.' I didn't want any trouble round this steel plate. But he still ached to have a go and finally got real mad. As we say in the Army he 'shoved his stripes under my nose' and told me to hand him the rifle, 'and that's an order,' he says.

---

20 Probably Gunner 3513 Stanley Pearson and Bombardier 195 William Herbert Felstead, 1st Field Artillery Brigade, AIF. See Appendix II, Notes on individuals mentioned in the text.

CHAPTER 22

So I handed over, deeming it best to say no more, and stood alongside him pissing myself with anxiety while he took a long and careful sight. Then there was a 'plop' and when the Corporal and I picked him up he was as dead as ever a man can be. Then the Corporal cussed and swore — said 'I'll get the bastard who did that to my pal' and in spite of my protests, poked the rifle through the hole himself. I hollared for our Sergeant, who came quick. But not quick enough and the Corporal, badly wounded, was later carted off by the stretcher-bearers. The body of the Staff Sergeant was carried to our front trench and we sent word back to his Battery as to what had happened.

Sorry, you two Australian Mums. Maybe I ought to have clouted them both to stop them being so damned stupid. But striking a superior officer while on Active Service is a deadly offence. Besides, they were both much bigger than me.

Some two hours later, three friends of the dead Staff Sergeant came to carry his body away. As I'd finished my two hours sniping duty, and feeling a bit responsible for the whole business, I offered to make the fourth man and between us we carted him away with the laudable intention of carrying him to an improvised cemetery that had just been started, about a mile away. It was a very hot afternoon and after some 200 yards or so over this rough and broken country, one of the Aussies who had been the Staff Sergeant's closest friend back home, decided 'one place is as good as another once you're gorn,' so in a tuck of the ground sheltered from rifle fire we laid our burden down and scratched out a shallow grave. A good hour's sweating work it was, too, and punctuated all the time by homely Aussie expressions of grief and regret, and maledictions on the hot sun that scorched our labouring bodies. Alongside us, and indifferent to the proceedings, the cause of all our labours lay stiffly, his wide-open eyes coldly watching us with the baleful malevolence of the dead. Whenever I looked at him I seemed to detect in those immobile marble eyes, a frightening malignity.

> An orphan's curse would drag to Hell
> A spirit from on High,
> But, oh! more horrible than that
> Is the curse in a dead man's eye.

In between his labours of digging, the dead man's pal continued with his lamentations. 'How I'm going to face old Jack's Mum back in Sydney, I just don't know. Her last words to me, Tommy afore we left, were "Keep an eye on him, won't you. He's a bit headstrong. Promise me you'll keep an eye on him".' He gave a deep sigh and rested from his digging for a time, squatting on his haunches and regarding his dead pal. 'You always was a cocky bastard, wasn't you, Jack?' he said kindly, and wagged his head reproachfully.

It was a macabre scene — this bare-headed, big, live Aussie — his sun-browned torso glistening with sweat, squatting alongside his indifferent dead pal and carrying on a reminiscent monologue of old times together. 'Remember that sheep farmer, Jack, who tried to gyp us of the week's pay he owed us for shearing, and the way you held him upside down to empty his pockets. And then the way you chucked him into the sheep-dip trough after, I thought I'd 'a died laughing.' Chuckling reminiscences like this went on for a few minutes — and I sat and watched 'em — an interested spectator of these two Aussies, the live and the dead, in comradely communion. Somehow, death doesn't seem to have the same dreadful sad finality among the irreligious soldiery as it does with their more religious brethren. It's more as if one of their gang had been posted away on detached duty. Maybe it's this attitude that enables them to survive, while still keeping their sanity, the daily harvest of the grim Reaper. Or maybe it is that in the blind ignorance of the teachings of religion itself, they've stumbled accidentally on the real truth of the matter. I dunno.

Then I blurted out what I considered to be my own share of responsibility for this tragedy. But the dead man's pal laughed this aside. 'You don't think, do you Tommy, that an Austry-lian boy is going to let any Pommie bastard stop him doing what he wants to do.' Switching suddenly to serious and solemn vein, 'Well, I guess we'd better do old Jack the honours,' and we laid him in his shallow grave.

Then the Aussie unstrapped his wide-brimmed feathered slouch hat from the top of his pack, straightened out some of the creases and knocked a cloud of dust off it before solemnly donning it, stood alongside the grave for a few seconds, then just as solemnly doffed it. The proprieties of interment thus duly observed and completed to his satisfaction, he said, 'I suppose someone should say a bit of a prayer. Any of you blokes know any — I've forgotten all mine.' There was a shuffle of denial from the other two Aussies, so I volunteered the Lord's Prayer, wondering at the time how many others from this place had gone to meet Him, 'which art in Heaven,' and armed with no more letters of introduction than this brief prayer. Then we filled in the space above the dead Aussie, and yet another hump appeared on this sacred ground of lost endeavour.

The other two Aussies were mostly silent, and didn't say much except for an occasional interjection of the soldier's favourite obituary, 'Poor Bastard.'

The dead man's pal raised the point of marking the grave with something — 'a bit of a cross or something.' I thought of all the other 'little bits of crosses' scattered all over the Peninsula in all sorts of odd places — little packets of soldier bones clamouring for immortality by means of little wooden crosses, crudely fashioned from pieces of biscuit box wood, and with their identities scrawled in indelible pencil that the rains had by now smeared into indecipherable smudges, and said, 'Did it matter much?'

But this great big hunk of sunburnt Australian manhood was genuinely and sentimentally shocked. 'Did it matter much? There's a little lady back in Sydney, Tommy, I've got to face when I get back. How can I tell her her boy's in an unmarked grave?' He seemed appalled at my callousness. One of the other Aussies begged a bit of wood from a cook-house some distance away and returned in triumph, and with our bayonets we cut some sticks of wood and with the aid of a spare bootlace, fashioned a rough cross and pegged out our dead man's claim to immortality. With much licking of the stub of an indelible pencil he recorded the name of the dead Staff Sergeant and stood back to admire his handiwork. 'That's better, old-timer. Now I can tell yer Ma you're all tucked up nice and comfy. How many 'l's in killed, Tommy?' I told him; also, told him how to spell 'action.' As regards the date, we hazarded a guess, but the Aussie thought accuracy was of first importance, and after canvassing the opinion of a few chaps idly watching we split the difference and settled for July 10th, 1915. Even now, the Aussie wasn't completely satisfied. He'd still got some more of the stub of the pencil left and the urge for registering immortality was strong upon him.

'Shouldn't there be some other words after the name, Tommy? A sort of good wish?'

'Requiescat in pace,' I suggested.

'Come again, Tommy?' said the Aussie, puzzled. 'What's them letters they put after a bloke's name — when he's dead?'

'RIP?'

'Yes, that's it, Tommy, What's it mean?'

'Oh, about the same, I guess.'

So Staff Sergeant Ballantyne, through an unfortunate error in calligraphy, became Staff Sergeant BALLANTYNE-RIP, which could cause all sorts of passport delays at the Final Frontier. Fortunately, the Feathered Frontier Guards in Shining White are well used to the vagaries of the Anglo-Saxon race and no doubt all would be put right eventually.

One of the Aussies had some fags and we sat around a bit and smoked. Oh! blessed and mighty rare weed — how it mellowed us. Discussions ranged about the future life of Staff Sergeant Ballantyne-Rip. 'Wonder what it's like up there,' said the talkative Aussie, waving an arm vaguely skywards.

'Rum sort of place, from all I gather,' one of the others said. 'Sitting round all day singing bleeding hymns and playing harps. Praising God and all that mush.'

'What do you think, Tommy?' the talkative one asked me.

Startled, I had to admit that I didn't knew much about it, but Sunday School teaching memories seemed to corroborate the description already given. 'Much about like your mate says,' I opined.

The Aussie appeared troubled. 'No Sheilas?' he asked.

'Sheilas?'

'Girls. Bints,' the Aussie explained impatiently. 'Ain't there none there? And what about booze and fags? Ain't there none there, too?'

I said I didn't know, but I thought not. Memories of childhood Band of Hope Temperance Meetings were still strongly with me, with drink as the Devil's Handmaiden.

The Aussie was pensive for a bit, then gave a long low chuckle. 'No Sheilas, no booze, no fags. Cor! I can just see old Jack's face when he finds out.' Then he corrected his unseemly merriment and said piously, 'Poor old Jack. And as nice a bloke as ever you could wish to meet.'

LOST ENDEAVOUR

PART III

(A hotch-potch of Gallipoli memories)

CHAPTER 23

Life is A Maze

The scorching heat of the summer, the growing sense of frustration at our lack of progress, the endless monotony of the sameness of life, the sameness of the same dreary food — they all combine to produce in us a grand lassitude. Life, it seemed, had come to a big Full Stop. We had reached the ultimate in supreme staleness. The novelty had worn off — the excitement of the campaign — they'd all worn very thin. Then out of the blue came a welcome respite. At 24 hours' notice, and in batches, our Battalion was taken off the Peninsula for a few days rest on the little island of Imbros a few miles off shore and completely out of shell range, almost completely out of the noise of battle itself. On the Peninsula we were never completely out of rifle range — even during our short periods of so-called rest on the beach. So this temporary immunity from rifle and shell fire was something unique for us and oh! so gratefully appreciated. To be able to move again and without the sixth sense forever alerted telling you when to duck for cover — it took us a few hours to get used to this new security.

We had embarked for this island paradise in little minesweepers and all sorts of odd craft, packing them so tight you'd think they would have keeled over. An early fatigue party had landed there two or three days earlier and erected tents and we enjoyed the luxuries of camp life. Nothing to do all day except sleep, rest, swim, clean up and forget that there ever was such a damned thing in the world as war. Not even a parade everyday — just a perfunctory rifle inspection, more I think to remind us that we were still in the Army and not at a holiday camp.

CHAPTER 23

But everybody left us alone. We had no money — not that this mattered much — there wasn't a damned thing you could buy — no Canteens or anything. But we had the quiet sandy sea-shore to lie about on, soaking up the warm Mediterranean sun — too lazy even to talk much. And there was always the calm Mediterranean Lake to caress our naked bodies. This magnanimous Mediterranean water, understanding all, forgiving all. For countless centuries this mother of all the seas had soothed and succoured with its maternal caress the fretful bodies of other untold legions. Warm as the handclasp of a friend, smooth as a virgin's kiss.

As far as we could ascertain, our part of the island was almost completely uninhabited except for two or three peasant families of Greeks, and with whom we were strictly forbidden to contact or disturb. But at times we'd watch one particular family at a distance. There was one old man with a high-prowed and gaily painted boat, propelled by a pair of outsize oars and a small sail, who carried on his daily task of fishing just off shore, returning to his family just after noonday. His family appeared to consist of his old wife and two queenly looking young women who occasionally during the day ambled gracefully to a nearby well, water pots on head, long flowing skirts reaching down to their bare toes. We spent many an hour watching this microcosm of peace-time life, the pastoral simplicity of the Greek peasant in his daily round, just doing what their forefathers had done for hundreds of years, leading their own simple lives. A haven of peace and sanity in a welter of war.

The special time of watching grew to be the hours when the two graceful young women floated off on their queenly walk to the well. It got to be almost a parade with us — we'd assemble on a little hillock just to watch them.

'A penny for thi thoughts, Sarge,' said one, interrupting a long and lingering gaze of our Sergeant at these two graceful beings about their daily task at the well. 'Wot's tha thinkin' about, Sarge, or does tha want me to guess —'

The Sergeant gave a deep sigh, then grinned. In one of his rare flashes of humour he replied, 'Ee, Ah dunno. But Ah'll tell thee this, if tha's thinkin wot Ah'm thinkin', tha deserves to 'ave thi face slapped!'

The few days rest went by like a flash and we re-embarked in the little boats, back to the Peninsula — re-embarked grudgingly, resentfully and sadly. No doubt the rest had done us a world of good physically, but psychologically we had to start all over again, get used once more to 'battle, murder and sudden death.' Except that this time it was harder for us, we knew now what we were going back to, whereas on our first landing on the Peninsula ignorance was bliss and there was the excitement of the unknown. For the first time of this campaign I began to hate the Peninsula and all that it stood for — I guess this few days rest and absence from it all had spoiled me. But within two or three days life resumed its old pattern and we even forgot we'd ever had a brief glimpse of heaven. In any case there was a big offensive looming up and that stops a chap from thinking too much about the past — all now is the problematical future.

The ever-increasing heat of the summer played the very devil with our spirits, and biscuits and bully-beef is not the ideal daily summer diet, especially if it's washed down with nothing more exciting than lukewarm drinking water, heavily chlorinated. Amazingly enough, however, there were no noticeable adverse physical effects, apart from a distinct lassitude in our movements and our general lack-lustre appearance.

The last offensive on about August 6th turned out to be the last big determined offensive on Cape Helles. It was just as abortive, for with the guns and equipment we had and with the enemy in their favourable and almost impregnable position, nothing short of an unacceptable casualty bill would ever give us real progress. The terrain sloped evenly and gradually up to the heights of Achi Baba, that dominating hill-top we gazed on so longingly and wondered if we should ever eventually capture it. If we did, the way to Constantinople would be wide open, for it was said the terrain sloped just as surely and evenly the other side. I'll admit I'm no tactician but it stuck out a mile that whoever held that dominating hilltop held the key to victory. The Turks must have known this as well as we did and had prepared their defences cleverly and thoroughly, and from their point of vantage could view our preparations for attack with comparative impunity.

Every movement of ours was under their constant and critical scrutiny, and whereas their own line of communication was by their all-land route, unobserved and with comparative ease of supply, but with us every darned thing we had must needs be manhandled from ship to beach and from beach to front line. Aerial observation on both sides was almost non-existent — daily a lone aircraft from either side would make a desultory flight, but those were the days when aerial observation, aerial photography and its proper interpretation was an infant art.

It would be churlish, too, not to pay tribute to the very determined and stubborn bravery of the Turks themselves in their resistance, and this factor alone contributed in no small part to our lack of success. All honour to a brave and determined foe. All I know is I'd rather have them for an ally than an enemy.

A week or so before this last offensive opened, our machine gun team had been moved to a well-camouflaged position in the broken country, a position designed to give maximum enfilading fire on the Turkish trenches when the push started, and to support our own troops in their advance. It was essential not to disclose our position before the advance, so except for an odd individual sighting shot on the enemy trenches we lay craftily hidden. We had enough rations for a week, but water was a thing to find somewhere. Because of our need for concealment there was to be no daily trip back to our lines to replenish the water, at least not in the daytime, but if we wanted to, we could replenish at night. After a few hours, however, we discovered a water-hole of fresh spring water and where the water bubbled out clear as crystal. What a welcome change is fresh clean water after the chlorinated stuff we had got used to, carted up from the beach in 20 gallon fantasies on the backs of mules.[21] No wonder we accepted the daily risk of the Turkish sniper covering this water-hole, instead of making the nightly trip back to our lines for our daily needs.

It grew to be a sort of 'Turkish Roulette,' these daily dashes to the water-hole — an exhilarating pitting your wits and skill against the sniper. A quick dash across the skyline, then lying doggo for a bit, then another short dash, another dodge to shelter, till eventually the water-hole was reached and half a dozen water bottles filled. Then the trip back — not so easy this time with the filled water bottles flapping

---

21 Watkins is probably referring to *fanatis* — metal tanks for carrying water.

against your body, and a vicious bullet thudding into the ground a foot or so away, and always the triumphant return. Three or four trips a day to this water-hole each of us taking turn. Damn silly, probably, taking these chances when we could in comparative safety have got our own water from our own lines at night, but I guess we thought it was worth it. Worth it, not so much because of the flavour of the clean spring water, but also worth it because of the daily excitement, the gamble with death and all of us afraid to be thought 'chicken.' When it was the turn of you yourself, you got all the excitement you wanted. When it was the turn of the other chap you got too another sort of excitement, gnawing your finger nails with anxiety wondering if he'd make it safely. A side-effect of all this daft behaviour of ours was that it forged even stronger the already strong bond of attachment you get in a small machine-gun team; your mate's danger was your own danger, his safe return, your own safe return. You don't get this same sort of close attachment away from the machine-gun team — say, when on normal Company duties, where it's more impersonal.

Our Machine Gun Sergeant frowned on our activities, but give him his due, he never dodged his turn in these silly expeditions. 'We didn't ought to be doing this bloody silly thing,' he said. 'Ee, but Ah've got a hell of a thirst — this weather Ah could drink a bucket full. 'Ere, Cobber, it's thy turn, isn't it?'

So the 'Boy Scout' element in my character raises its ugly head. 'Nothing to it,' I'd say nonchalantly, 'Gimme the bloody bottles.'

'Well, watch thi step, Cobber. Ah don't aim to get no VCs if tha gets hit.'

Keyed up, apprehensive, but excited, I make the first dash over the skyline and beat the bullet that nearly parts my hair by a split second. Flat down on the ground I recover my breath and ponder on the next move to get over the next bit of the skyline. My cap on the end of a stick attracts a quick bullet, then a quick dash before the next shot — and so the game goes on till I reach the water-hole. And then the same thing for the return journey and a bullet hole in my cap to proudly display to my mates.

As I said, young soldiers are always daft young buggers. I guess to us it felt a bit like playing cowboys and Indians like we did as kids. However, we survived and with no casualties, which was better than we deserved.

When the main attack opened, it seemed to coincide with an equally fierce enemy attack and for the next two days we had plenty to think about. Our rations were by now exhausted and our ammunition was very low, so someone had to run the gauntlet and find our own men to get some more. We cut the cards for it and I was the Joe Soap.

It was a blazing hot afternoon and the ground was thick with smoke and battle. Hardened campaigner as I was by then I was appalled at the immensity of it all. Bodies lay about all over the place, and none of us knew which side occupied which trenches, so I made a dash for the shelter of the nearest trench about a couple of hundred yards away. Hidden by the smoke of battle I was lucky to reach it unobserved and dropped exhausted into its grateful cover. The trench was empty, empty except for one figure a few yards away — one of our officers.[22]

The appearance of this particular officer shocked me. Unshaven, haggard and bloodshot-eyed, it was difficult to recognise in him the one who was under all conditions the soldier 'par excellence.' His legendary smartness had never seemed to desert him ever since we had landed, and in the heat of previous battles he was forever of a parade-ground smartness that was a source of wonder and inspiration to us. When we would be mucky, battle-stained, dirty and unkempt, his leggings were always of the same beautiful polish, his uniform impeccable, and his face always scrubbed and clean. Like an invincible and inviolate god, he always appeared to us — by contrast. Nor was he ever flurried or flustered like us — always calm, efficient and superior to all conditions, even appearing always slightly amused by everything — like as if this sort of soldiering was some sort of game.

All the greater therefore was the shock when I met him, like a man who had been on a three days boozing jag and hadn't yet got over it. No battle hazards in the world could have reduced this officer to such a state — he just wasn't that sort of chap. Then I remembered whispered rumours that had been going around about him — rumours whispered so discreetly and sorrowfully by us, for no man enjoyed such respect and admiration as did this chap. So I began to wonder what sort of Dear John letter he might have received to have reduced him to such evident wild despair.

---

22 See Appendix II, Notes on individuals mentioned in the text.

In our early days in Cairo we had always thought of him as a bachelor. Some said he was married, others thought differently. We were never quite sure; anyway, we weren't all that interested. If we had ever seriously considered the matter at all it might have been when we all seemed to agree with Cliff Stansfield's summing up during one lazy evening's desultory argument on the matter. We'd just seen this smart spick-and-span handsome bloke go past us while we were squatting with our backs to the barrack room wall, enjoying the last of the evening sun.

'No, don't get up boys, sit where you are,' said this officer, flashing us one of his famous friendly smiles as he passed near us on the barrack square. 'Good night, lads, good night,' — and he disappeared into his quarters.

'Don't 'e ever go out, like t'other officers,' one of us said.

'Most times, no,' said another, "e seems suited, just as 'e is. Besides, some of t'other officers 'as got their missuses in Cairo. Anyway, maybe this one's not married.'

'Oh! Ah dunno. He could be married. Ah'll find out sometime from 'is batman. Ah never thought to ask 'im.'

'Course 'e ain't married,' said Cliff Stansfield decisively, and joining in the argument. 'Why! 'e don't even *look* married,' asserted Cliff, reasonably enough. 'He looks 'appy.'

But shortly after we heard that this officer was, in fact, married and that his wife had sailed from England and just joined her husband in Cairo.

Then after a few weeks we noticed a change in this officer. He no longer seemed to bubble with the joy of life as heretofore. There were one or two sly and completely insignificant rumours in the Wet Canteen that his wife was thoroughly living it up in Cairo. But it was no more than a rumour — just a mere wisp of a rumour, of the sort that usually floats around with soldiers, and which passed almost unheedingly by us as a mere bit of salacious gossip, hardly worth repeating. Then, and some two or three months after we had been on Gallipoli, one or two chaps returning from sick leave in Cairo, brought back stories which disturbed us. Saddened us, too, because we loved this smart, one-hundred-percent keen soldier, spick-and span, 'above-it-all' officer who was such an inspiration to us by his very appearance and martial behaviour. The beau-ideal of an officer, this chap, a very 'chevalier sans peur et sans reproche.' Elegant in speech and manner, handsome, debonair, a keen disciplinarian, who used to dish out to us our well-deserved punishments in Cairo — dish them out impersonally, strictly, but somehow almost apologetically. Maybe it was his twinkling eye and the way his well-clipped moustache seemed to hide the suspicion of a smile — an understanding and sympathetic kindly smile — that took the sting out of his punishments. I dunno. But young soldiers are quick to tune in to a friendly nature, so to be weighed-off and punished by this chap was something we even liked. But don't ask me to explain the intricacies and contradictions of human nature. Maybe, too, we respected those campaign ribbons across his tunic. All I do know for certain is that there wasn't a man of us who wouldn't have gone to the ends of the Earth for him.

And then when one of your own mates, back from a brief sojourn in a Cairo Military Hospital after a flesh wound, starts telling you, and a bit regretfully, in a quiet trench corner, things you sort of wish he'd never opened his mouth about, but curiosity compels you to stick around and hear more. Like how, the day after the Battalion had left Egypt and sailed to Gallipoli, his wife had started to really go the pace with the other officers of the Cairo garrison left behind. 'Ay, it gets some women that way, mate, does Cairo, sometimes. Reckon it's the heat, and being married and all that… if y'know what Ah mean…' and grins. 'Makes 'em act funny — like they'd never act in England, if you follow me. The heat, Ah s'pose, and all them Officers Mess Dances. Cor! I remember one o' them dances. On Bar Duty, I was — it was all right, too, free drinks for me all night. But some of them officers' ladies! Cor! the things I've seen at these dances. Goes to their 'eads, all of 'em, Ah reckon. The uniform and all that. An' bein' away from 'ome. An' the heat. But this one — (the officer's wife) — she wur the talk of Cairo.'

'Does he know, then?'

'Well, Ah reckon not. And Ah wouldn't like to be the one wot told 'im. Not me, mate.'

Maybe this unwitting conspiracy of silence was easy. Egypt wasn't more than a few hundred miles from Gallipoli, but it was thousands of miles away in essence. There weren't enough leagues in the universe to divide our present surroundings from the lazy peaceful atmosphere of Cairo.

So God alone knows what had happened to effect this metamorphosis in this so popular officer. One thing's for sure — no battle conditions could have shattered him so. He wasn't that sort. Now if it had been me... well, every rattle of shots ruffled my feathers quite noticeably. But not this chap.

He crowded alongside me in the narrow and unoccupied trench, where instinct made us keep our heads down and crouch together. He asked me what I was doing here. I told him, but it was like talking to someone miles away. And I had the horrible feeling that he knew what I had been thinking about — about his wife, I mean — and it made me feel more than ever uncomfortable. Then he said that he himself would venture across the broken terrain and try to establish contact with the battalion, but I demurred and said it would be better if I did — 'me being less important Sir, than you, if anything happens — if you know what I mean, Sir.'

'Less important!' he said in a cracked voice, and with a hollow laugh that must have come up straight from Hell. Then there was a momentary flash in his eyes of the old officer I had always known — the familiar kindly smile flickered under his moustache. He gnawed his lip a bit, and the glare in his eyes had me scared. 'Less important! So you think so, do you, Watkins?' and eyed me intently. And I blundered on, 'Well, Sir, you look all buggered up, Sir, if you don't mind me saying so, so I thought maybe I could make it quicker.' He said nothing for a bit, then, 'So I look all buggered up, do I — maybe I am, what you say lad, all buggered up,' and that gleam in his eyes was the look of a man who had just shaken hands with the Devil. I felt damned uneasy. His twitching face, hoarse voice and wild demeanour had me worried quite a bit and I'd have been glad to have jumped out of that trench there and then and try to find the rest of the battalion. 'All buggered up, as you say, Watkins, all buggered up. Maybe I am. But what's it matter, eh? What's it matter?'

'Look, Sir, if I dash smartly across to that trench when the shelling eases a bit, I could...'

But with a touch of his old masterful and imperious manner, he said 'Stay where you are. I'll go — and that's an order.' And he climbs slowly out of the trench and starts to walk. Walk, mind you, not run, towards the trench we could see about 400 yards away.

Soldiers quite often get second sight. I knew — just as sure as I know my own name — that it would be his last walk.

Had he left it to me, ten to one I could have made it. But don't get any ideas of heroism on my part. Nearly 20 years younger than he, there would have been no abnormal risk for me. At least, me being what I am, I'd be taking no undue chances, and I'd have been across that ground as fast as a scalded cat. A quick dash, a dodge to shelter, another quick dash, and if the luck of a soldier still stands I'd have made it. It would have been no more than we had been doing for sport at the water-hole this last few days. And I had the sublime cockiness of extreme youth — 'it could never happen to me' feeling.

But I guess he wanted it that way — to go out like that. Out of his troubles the soldier's way. The quiet matter-of-fact way, like the blokes of our race have a habit of doing sometimes. And which of us has the right to deny any man his own chosen exit? It's been done before. Nearly 2,000 years ago, in fact, and no doubt umpteen times before and since. Such blokes never really die.

'To turn one's back on the tyranny of Existence.' It needs a special sort of guts to do this.

There's times, too, when death may appear to be the ultimate stupidity — the final mockery of a life-long endeavour. And on the battlefield it is all too often a messy, bewildered and whimpering exit. But there's times, too, when this exit has a dignity all of its own. Such as when this officer deliberately forced his feet forward to grasp the hand of the Distasteful Dispenser of Blessed Oblivion... or to seek the Life Everlasting.

There's times when I think of Death, of this 'Life everlasting, the Communion of Saints' — and all that jazz. For me, and for most old soldiers, death has just got to be an Old Pals' Reunion, the chance to meet up again with the old buddies who left us so hurriedly on the battlefield. To meet again, not the Saints, but the ordinary coves, the little ordinary blokes whose sole claim to a seat on the clouds may be no more than that they did their duty blindly, faithfully and un-noticed. Surely in Heaven, there'll be a special little corner for scrubby little coves like us.

Vaguely I call to mind some lines I read many, many years ago, something about

> Life is a maze of many strange streets,
> And Death — the Market Place where all men meet.

If it's true, there'll be lots and lots of scruffy little soldiers like us dodging in and around the market-stalls, seeking out their mates of yester-year.

I waited till nightfall and eventually stumbled on my company, or what remained of it. There were some grim stories I was to hear.[23] It was all too evident our local push had been a disaster. In the comparative quiet of the next few days, I was to see in various spots of this ragged terrain the price paid for glory. Bodies lay about in clusters, in pairs, singly — and at all sorts of unexpected places where some of the wounded had dragged themselves off to die. And in a corner of that little Krithia vineyard, it was there a corporal and I came across the 6th Battalion, Manchester Regiment boys I have mentioned earlier. 'Outstanding in their neatness and prissiness,' I had referred to them and somehow they seemed extra neat as they lay there, packed in close array. We stared at these lads for some little time, saying 'nowt' — just looking. What the hell I was thinking about as I gazed on them I can't even remember. Maybe I was a bit shocked at first in seeing so many of them packed together in such a small place. I dunno — I honestly can't remember. It was the Corporal himself who pressed the right mental button to make the penny drop for me.

'Looks bloody queer, don't it,' he ruminated.

'What does?'

'All them lads like that. Looks queer, don't it, mate?'

I agreed. 'Bloody shame, too — nice lads like that,' I said.

The Corporal stared at me for a bit. Then — 'No, you don't catch on, do you, mate,' he said.

'Catch on?'

'Ay,' said the Corporal impatiently. 'Wot does ta see about 'em wot's different,' he asked me.

---

23 Battle of the Vineyard — Battle of 6/7 August at Helles. See Appendix III, Notes on events and places mentioned in the text.

'Well,' I said — rather stupidly perhaps, 'you mean — being all dead — like that?'

'Tha daft bugger — look again. Then tell me what tha sees is different,' the Corporal said angrily. 'Tell me wot's different about 'em, or does ta want a pair o' specs or summat?'

Then the penny dropped. They were all *properly dressed*. Properly dressed — not like all the other bodies we had come across that were garbed, not like all the other living ones, too, on this Peninsula, all garbed the same way in our usual and favourite 'scruff order.' Our usual 'scruff order' of grubby shirt, shorts, boots and socks and 'hats assorted', the very most of dress we could bear to don on these scorching hot days of summer. But these boys — neat khaki drill tunics, properly buttoned up to the neck, buttons polished, bayonet scabbards clean and honed, neat regulation, clean pith helmets. Even their belts and bandoliers were clean and polished.

The Corporal seemed temporarily bereft of speech. 'Well, I'll be buggered,' he burst out at last, 'Ah wonder who the officer wur who made 'em go into battle properly dressed like that.'

The Corporal's right. One wonders, too. Maybe one even shares some of the pathetic regimental pride of the officer who had decreed that these boys of his, due soon to be transferred to Higher Echelon, should do so properly and smartly dressed. I tried to put over this point of view to the Corporal, but he spat disgustedly.

'Tha allus was one of them 'quare fellahs' wasn't tha, Cobber,' he said. 'Tha allus thinks round bloody comers, doesn't tha, tha crazy bugger,' he added compassionately.

Then he tore his gaze away from the sight in front of him, half turned his back and puked a bit.

'Come on,' he said, 'let's move on, and before Ah spews me guts up looking at these poor bastards.'

And to avoid any further excursion into sentimentality, we took refuge in profanity. Profanity... and obscene vulgar abuse of 'all bloody silly officers,' and affecting a callousness of the whole business of war, generally. For in wartime, and unless you keep your sanity-saving callousness handy, unless you have it on tap on special occasions, you'd be in a nuthouse in no time.

Then we moved on, surveying the remainder of the result of these past few days of battle's bloody toll. We collected identification discs from some of the bodies we stumbled across so they could be properly registered as dead, but we got tired of it. There were so many and somehow it all seemed so fruitless. Bewildered and lacking guidance we toted these strings of identification discs, eventually trying to hand them in to a Sergeant. Equally bewildered and aghast at the immensity of it all, he told us to keep them for a time. I forget what happened to 'em. We probably chucked them away at last in a blind endeavour to erase our memory clean of all that had happened in the past few days, and gropingly began to gather up again the pieces of our normal life — things like pounding to a powder with the hafts of our bayonets the hard Army biscuits, and with a bit of melted fat and some jam childishly vying in competing in producing the best substitute for a jam pudding, and cooked in our little mess tin lids; things like arguing fiercely about the town's rugby team; things like — oh! any bloody thing so long as it wasn't war — in a desperate effort to hang on to your sanity. For folks are like that. You saw the same thing in World War II, such as just after a blitz when folks would go out of their way to cling on to their sanity by an emphasis on the domestic commonplaces of life — even to the point of banality.

This Gallipoli campaign was a queer sort of war — more like a test of valour than a properly organised 'will to win' war. Ignorant as we were of the proper conduct of war, we all felt instinctively the amateurish effort in its direction. The nature of the country, too, was such that concerted all-along-the-line action was either too difficult or too badly organised. I suspect the latter, but a private soldier can't be too sure. He doesn't see enough of the general pattern to really know what's going on but at times he makes some shrewd guesses. All I know is that heavy engagements went on in certain parts of the front line, and less than a mile away on our own front there could be comparative quiet and disinterest and with no-one taking the pressure off that part of the line heavily engaged. This sort of thing doesn't fit in to any pattern of war, and especially on such a small battlefield as ours. Knowing nothing then, as we did, of the proper tactical conduct of war, it didn't seem all that strange to us at the time. It is only when you get more experience of war that you realise what an amateurish 'do-it-yourself type cock-up the whole conduct of this campaign must have been. But from the time of this last big push on the Cape Helles front our complete lack of confidence in our superiors conducting the campaign made us cynical, despairing and savage.

We did our duty, but from now on we did it without hope, without enthusiasm. And that's a hell of a rot to set in, with any Army.

CHAPTER 24

If I Touch It with My Stick

Scattered all over our beloved country, in every town, city, village and hamlet, you'll find they all keep their own individual 'Book of the Dead,' those stone-carved crosses symbolising our future hope, and our present tribute to those who, according to their lights, tried to make this hope some sort of reality — lists of names that the stone-mason's craft has carefully inscribed with hammer and chisel, usually at the base of the Cross, and — in case you may have forgotten — three sparse, bald words reminding us that these blokes were 'killed in action.' Most of us haven't the vaguest idea what all this 'killed in action' stuff really signifies. If we think about it at all it merely suggests a quick and merciful execution by a stray bullet, a matter of a few seconds maybe. If it's more than that we just don't want to know.

'Their name liveth for evermore' — 'evermore' being the uncomfortable few minutes when, perhaps, we may glance at these names once a year. It's sufficient if we place our annual tribute of red-paper poppies on the steps at the base of the stone cross, hurriedly glance at one or two names carved thereon, and then just as hurriedly try to forget. It's not our business to wonder how exactly each and every one of them might have died, or under what particular circumstances anyone of them had 'handed in his checks.' Maybe it's just as well. What dramas, what heroisms, what agonies and what pathos might be revealed if only we did know how each and every one of them had died. Some deaths spectacularly heroic, some just as heroic but unspectacular — the little ordinary coves who—and quite uncomprehending what it was all about, did their duty just the same. In one of these In Memoriams (including my own town) appear the words –

> Those who fell did not choose to die. They were mostly in the flower of early manhood with their lives before them. We do not know how all of them died... but we are content to know that all did their duty, and would not have acted otherwise had they known their fate beforehand...

This thing called Duty! Sometimes I think this word has bedevilled mankind more than anything else since the Garden of Eden. Maybe it's the real 'original sin'—this 'original sin' that has so puzzled the Ecclesiastical Pundits ever since they first invented the words to excuse man's native genius for skull-duggery. But before I once again falter into the mumbling and inchoate philosophies inseparable from old age, let me tell you about how one bloke died.

A great bloke, our Battalion Captain Quartermaster. And this time there's no indiscretion in naming him—Captain William Henry Griffiths—South African War veteran, cool, caustic and clever.[24] We'd accompany him at times, three or four of us on a ration fatigue to the Stores on the beach and collect in sacks various rations for the Battalion. The stores etc. would be laid out in separate piles, with our Capt. QM, list in hand, directing us what to pick up and stick in our sacks. It was almost routine—we knew exactly what was due to us, and the variety couldn't have been more limiting. But there were special tit-bits lying about in neat and orderly piles—not meant for us—luxuries like tins of fish, and Blackwell's special jams, tins of butter, etc. But definitely not for issue at Battalion level. Brigade HQ perks, these, or for some special types of High-Ups in the officer line.

Our Capt. QM walked with a limp—legacy it was said of a South African war wound, and his walking stick as he limped his way through these piles of stores was as familiar to us as the cynical grin on his face. A familiar and potent magic stick, this walking stick of his. 'If I touch it with my stick, whip it up' he'd say, his eyes crinkling with grim

---

24 See Appendix II, Notes on individuals mentioned in the text.

humour. So as we wandered through these piles of stores collecting our legitimate entitlements under the watchful eye of the chief storeman, our Capt. QM would be waiting for the moment when the storeman's eye was momentarily distracted. A quick flick of the stick on some coveted pile, a bit of smart soldier legerdemain on our part, and a few bits of forbidden luxuries would join the legitimate entitlements in our sacks. Back at Battalion HQ — and to give the old Captain his due — he was never mean. We'd get rewarded for our official thieving by some tit-bit — say a tin of butter to share out, or something equally precious. 'And keep it out of sight, you thieving so-and-so's' and he'd grin like an Imp from Hades. So naturally there was no lack of volunteers for this particular fatigue duty.

On this fateful August day, I heard how he did his last duty. I wasn't there myself, I was with the Machine Gun team.

All the Company Officers had been killed or wounded in some Turkish counter-attack, the Company driven out of the newly won trench, beaten, thoroughly demoralised and ready to cut and run — like soldiers get sometimes in the heat of an engagement, when out of nowhere emerges from his non-combatant duties our Capt. QM, limp, walking stick, cynical grin and all. He surveys the thoroughly demoralised remains of the Company angrily. 'Ready to cut and run, are you, you buggers. Run from a lot of bloody Turks. Lancashire lads running away. Nay, never let it be said. Here, Corporal, help me over the top of the trench. Easy now, mind my gamey leg. Easy now. Ouch! Mind my leg, blast you. And now pass me my stick and follow me, all of yer.' And waving his walking stick high above his head he limped smartly towards the Turkish trench followed by the now heartened and electrified remains of the Company, angry and ashamed of the last few minutes. There was a brief and savage encounter and the trench was recaptured. Later they carried in the body of the Capt. QM, and some bloke got a slight flesh wound in the leg going back for the Captain's walking stick, for stick and Captain were inseparable. Both Captain and stick were later buried in an improvised grave, and the grave marked with a home-made cross of biscuit-box wood.

'Ee, but Ah wish Ah could a' kept that walking-stick. Ah do an' all,' said the bloke telling us the story of this incident. 'Sort o' souvenir, like. But Corporal said to bury it, like, with the owd chap. Said as 'appen he'd need it later on, like. Not that it made much sense to me, wot the Corporal said, but 'e wouldn't let me keep it.'

Maybe there wasn't a lot of sense in what the Corporal said. But for that matter, there's not a lot of sense in any battle.

Me being what I am I prefer to think of some 'up-Top' Quarter Master and His party of Marauding Angels hovering over the battlefield. Maybe He was a bit short of good replacements for His sky battalions and wanted a few good reinforcements — some 'extra-special' blokes. So maybe He too says 'Anyone I touch with My stick, gather him in. But I only want the best, mind you.' So maybe His stick touched our Captain QM.

Or maybe you yourselves can think of a better explanation.

* * *

A lonely, isolated figure, this Capt. QM of ours, during our training days in Egypt. Seldom would you see him hob-nobbing with the other officers. Maybe it's because he was so much older — about 50 I should guess — as against the 20 to 25 years average age of our other officers. Or maybe those ribbons on his tunic, or that legacy limp from a previous war singled him out as one different — one who knew the grimmer side of war. Or maybe the very nature of his regimental duties as Capt. QM kept him aloof and pre-occupied — kept him out of the joyous enthusiasm of the younger and war-ignorant exuberance of the nice young chaps who comprised the general run of our officers. These young boy-officers, they were all more or less of the same social background, sons of solicitors, sons of mill-owners, sons of men who lived in fine houses and had plenty of 'brass' — and maybe our Captain QM was not out of the same top drawer, socially. Not that there was any lack of respect for him — quite the contrary. After all, a beribboned tunic and a legacy of a battle limp are not things that soldiers can treat with disrespect — especially young and inexperienced boy-soldiers like us who had the usual awe for the old campaigner.

Every day in Egypt our pattern of training was the same. A long gruelling march in the desert, under the hot sun, to build up our stamina. Then open-order skirmishing, each section in turn dashing forward some 50 to 100 yards, and their advance covered by the supporting fire of the other sections — supporting fire of lots of exhilarating blank cartridges being fired madly by us. Then finally a bugle call, a fixing of nice, well-polished and shiny bayonets on the end of our rifles, and a dashing charge of glory-mad young boy-soldiers towards a mythical and non-existent enemy, who presumably fled terrified, at the very sight of the legendary British bayonet.

Such was the stuff of war on which we were trained.

Then the bugle would sound the recall and we'd gather together again — like children do after an exciting picnic — 'tired but so happy' — and take about half-an-hour's rest before setting out for the tiring march back to barracks. Half an hour's rest, smoking innumerable fags, flushed with victory over our non-existent enemy, swapping stories we'd heard of the might of the British Army — the terror of the enemy at the sight of the British bayonet. This Capt. QM would move amongst us, mostly 'not saying nowt' — his eyes crinkling in amusement — a solitary figure — like a fond old father — and picking his way through the hundreds of reclining figures on the sand, and eyeing us amusedly. Occasionally he'd stop to exchange a few words with some of us. Sometimes gently sarcastic such as 'be sure to wipe the blood off your bayonets, lads' — and then roar with laughter. Then sometimes genuinely kindly. To a bare-headed lad sitting in the mid-day sun 'keep your helmet on lad — sunstroke's a bad thing to get.' But we always felt a bit uncomfortable when he was around… like schoolboys who had been too exuberant in their play.

But essentially a kindly man. Once, crossing the Citadel barrack square, passing him and forgetting to salute, he calls me back. Gazing above my head into space he said 'Watkins, when I was a young soldier, we had a damned bad habit of saluting an officer when we passed him.' Confused and ashamed, for I was a keen young soldier, I apologised, saying my thoughts were miles away. Then he takes the sting out of his reproof, enquires after my Dad, whom he knew, and in seconds we were chatting like old buddies. That's the best of a Territorial regiment — it's just a big family unit.

The man who disbanded these Territorial regiments did this country a great dis-service.

Why do I natter on like this — like a gossipy old woman? I'm damned if I know. Honestly, I just don't know. Maybe it's the pleasure I get in reviving the memory of a decent old-timer like this chap, who 'took his discharge' on that fateful August day — took it honourably like a good Lancashire Officer, waving his stick above his head and limping towards the enemy in front of his men. This country will always need men of his pattern. Let's hope we haven't lost the mould.

CHAPTER 25

Mr and Mrs Fly

The last six weeks of the Summer still register with me, and after some 50 odd years, as the most wearisome, the most sordid of the whole campaign. Casualties were few after the August push had abated, for the pace of the struggle weakened. Or maybe the Turks were suffering from the same thing as us.

A gigantic plague of flies. It is difficult to believe it was confined solely to our army. Without doubt the Turks were doing the same thing as us, crawling about listlessly, weak as kittens, victims of a violent diarrhoea that never left off even for an hour. In any country there's always an invasion of hordes of flies at the end of a long hot Summer. Not too difficult to endure, these pestilential plagues, under civilised and hygienic conditions, but hygiene was a word we hardly know how to spell. Our open-type latrines, the thousands of opened and still full jam-tins lying around as the despairing soldiery searched these falsely labelled tins, only to find the abhorrent 'Plum and Apple' coyly nestling under another and more attractive name.

Tins and tins of jam, hundreds and hundreds of 'em, opened monotonously and endlessly in the fruitless search for just one tin of El Dorado, and just as monotonously and endlessly tossed away over the trench top. A real field day for the flies. No rolling snowball ever grew so quickly as did this plague of flies. It was a physical impossibility to eat anything in the daytime — impossible to dodge anything quickly into your mouth without spitting out at the same time the flies who insisted on hovering on your biscuits, your bully-beef, crawling over your lips, your eyes, and searching every inch of your face.

Then our unhygienic latrines — the open 'bog-hole' round the trench corner. The bloody flies — straight from Hell these bloody flies. While you squatted over a hole, they'd swarm and crawl over your bare posterior, while down below Mr and Mrs Fly feasted cosily on the excreta.

'Quick', says Mr Fly to his spouse, 'there's a soldier just going to eat a biscuit. Let's hurry before he finishes it.' But Mrs Fly, fastidiously brought up from childhood, and hesitant — 'Hadn't we better — er — wipe our feet first — you know...' And Mr Fly, petulant and scornful 'Oh! What's it matter — these soldier boys — they're used to it — it can't hurt them.'

So no wonder we resisted the pangs of hunger during the daytime, and crawled to the nearest latrine hole in the comparative cooler night and when the flies weren't about. And sit there under the stars, blanket round our shoulders and munch our dry biscuits, only to feel them pass through us within minutes. Too weak even to scrape enough earth together to cover what we had deposited, we'd totter back to our trench again, grateful if only we could get just one hour's respite before another call to the latrine. Weak as starved kittens, hollow-cheeked, lack-lustre eyes, and losing stones in weight, we endured this for 4 to 6 weeks. A troop of determined Brownies or Wolf-Cubs could have captured our trench without a fight.

Then the first welcome, God-given frost. I never thought I'd ever welcome cold weather so gratefully, for I'm the type that tends to whimper a bit in cold weather. A keen frost killing all the flies — a frost that racked our emaciated bodies and rattled our very bones — but a frost oh! so gratefully welcomed. So bit by bit we recovered our drained vitality, but it took time. And hardly before we were ready came the Winter — bitter cold — this Crimean-type winter we were to endure, the like of which, thank goodness, I have never again encountered nor ever want to anymore.

CHAPTER 26

This Brotherhood Lark

'The best way to stop a dog-fight,' said the Lord God Most High, 'is a well-aimed bucket of cold water.' Which is probably why one day in late November, He ordered all the buckets for miles around to be filled.

All morning we'd felt the nearness of the approaching storm. By about 3 p.m., the sky started to blacken. There was an uneasy electric and irritating tension in the air. In any country there's all sorts of queer goings-on just before a violent storm. In England horses race madly round the field, cows lash out clumsily at one another. The cat races upstairs to hide under the bed and Fido whimpers in his corner.

But English storms are mere zephyrs compared with some of the storms you get in these Mediterranean countries. If ever you've experienced one — say, while on a holiday in Spain or some place near — you'll all be familiar with the high state of apprehension and irritability that develops just before one of these Mediterranean storms. The Spanish Senora hurries the gleaming cutlery off the table into a nearby drawer and covers it with a thick cloth, and flutters round the room straightening the little crucifixes and sacred statues that hitherto have preserved them from a 'sudden and 'unprovided death' by lightning and thunderbolt. She glances nervously at the Senor of the household, who lordly pooh-poohs her fears — and privately wishes he hadn't dodged going to Holy Mass last Sunday, and makes a silent vow to say three Novenas once they are safely delivered from the oncoming storm. They drop the latticed blinds and bolt all doors to keep out the thunder and lightning, and El Senor worries a bit about the loose tiles on the roof and wishes he had got them fixed last week — as he meant to, but didn't get round to it. The gathering wind rattles the doors until the whole place is a pandemonium of noise and the windows shake so much you'd think it was the Devil himself trying to force an entry.

In our open trench we had been working off our own high tensions by starting a fusillade of rifle and machine-gun fire, and — maybe under the influence of the same tensions — the big guns farther back started their own brand of activity. No less wound up and equally highly strung, Johnny Turk retaliates, and there's such an infernal bloody row, you'd think a big offensive was imminent. But there's no need to worry, brothers — it's only the electrical content of the unkindly heavens playing bloody hell with our nervous systems. We didn't have a lot of cutlery to clear off our non-existent tables but our Platoon Sergeant, ever on the ball as regards the practicalities of war, told us to sheath our bayonets — 'steel attracts lightning, tha knows, and it's best to be safe, like.'

'But wot abart mi rifle, Sarge — t'aint made o' wood, tha knows, Sarge.'

'Ay — but that's a chance tha mun take, lad, tha can't let go o' that.'

With this dubious combination of scientific foresight and military necessity we had to be content. We didn't have any latticed blinds to draw down, but we cloaked up in our ground-sheets, ready for the deluge. By now it was as black as night. 'Like the end of the bloody world,' said one of our chaps, frightened, and licking his lips nervously. It certainly was eerie. An occasional flicker of lightning warned us it was getting near to 'curtains up' for the big show.

Nothing holds for ever — not even an imminent storm, and the Cosmic bombardment opened up with a continuous roar of thunder so loud you'd think it would blast the whole world apart. The lightning flashed and played in one continuous whole. The sound even of our own big guns' bombardments was dwarfed into puny insignificance compared with what came out of the skies, and as if realising the futility of this competition in noise, all the guns on both sides became silent and gave up. Then the sluice gates of Heaven opened and the cold rain fell drenchingly, not in drops but literally in buckets. In less than a minute the floor of the trench was a sea of mud, ankle deep. Ankle deep mud was not strange to us but it was with real alarm we saw the trench itself filling fast with water. Soon it was knee deep and we climbed on to the firing step to avoid it. But it still rose higher and higher until the firing step itself disappeared from view. Little tucks in the side of the ground in the front of our trench became deep crevices under the weight of the falling rain, and the water started to stream in through them, in countless little rivers. By now the water was a good three feet high and rising fast as it was fed still more by the water swirling down from the trenches on the higher ground.

'Wot the bloody 'ell do we do now, Sarge,' somebody called out, panic stricken. The Sergeant was plainly baffled. Such a contingency as this was never even referred to in any Manual of Military Training. But the issue, for once, was decided for him. 'Ah'm not stoppin' 'ere to be bloody well drown-ded. Ah'd rather stop a bullet' — and one of our chaps climbed up into the open in front of the trench. Quickly we all followed suit, first throwing over the top of the trench such possessions as we had managed to hang on to during this flooding. Anyway, a bullet couldn't make our misery much worse.

'As Allah wills,' says Johnny Turk. 'Perchance I'll find it better to enter Paradise by a bullet from the Idolatrous Christian dogs — may Allah, the All-Merciful, the All-Compassionate, rake their entrails with their own barbed wire. Better to go to Paradise with an idolatrous bullet than to stay here to be drowned like a rat.' And he, too, climbs out into the open in front of his own trench.

And soon the whole line of front trenches is a line of British and Turkish soldiers, stood out in the open in the enforced companionship of over-riding and mutual misery.[25] Adversity sure makes strange bedfellows. By a miracle — or by some fast field-telephone work — all the big guns behind us and in front of us remained silent. For either the British or Turkish big guns to have opened fire would have meant mowing down their own men too.

Then began for me — and for all of us for that matter — my first lesson in the all-embracing stupidity of war. There are times when I search my unlettered mind for a better word, but I guess that even with a University education I'd never find a better word than *stupidity*.

The lesson we learned we soon forgot — after the pattern of mankind since the world began. I guess that not more than 0.00001 of one percent

---

25 See Appendix III, Notes on events and places mentioned in the text.

of lessons learned throughout the ages have ever been remembered by man. Man is quick to learn — but forgets a damn sight quicker.

As far as your eye could see in the bad visibility of the torrential rain — to the right, to the left, and in front were the long lines of shivering wretched Tommies and Abduls, facing one another in the open. Curiosity being one of man's characteristics, we gradually and ever so gingerly edged towards one another. At first a bit warily, like dogs of opposing packs, suspiciously, all our hackles rising and bristling — and stiff legged like the two lines of Western gunmen you see sometimes at the Movies, getting ready to shoot it out. But all we ourselves, Turks and British, wanted to exchange was mutual misery, not gun-shots.

The British Tommy is the friendliest bloke in all the world, given half a chance, and then began the most monumental fraternisation of Turk and British — dead unethical, but dead human. Astonished, both sides discovered — and as if for the first time — that the others were human beings like themselves, two arms, two legs, eyes and nose exactly the same. We even shared the same sort of sodden, water-logged uniforms, we even shared the same woe-begone air of drowned misery. There was more hand-shaking and back-slapping than there has been in this part of the world since the beginning of Time. Like excited monkeys both sides tried to establish a bridge of communication — us in our pidgin Arabic picked up in the sleazy bars and cafes of Cairo, and some of them paying us the compliment in courteously responding in their own brand of our fabulously ancient and cultured Olde English tongue. Truly British, we opened the conversation with the weather. With expressive gestures skywards, we commented, 'Musquise, Abdul, Musquise.' They agreed feelingly. 'Ai! Ai! Tommee — no bloodee good, no bloodee good,' and flashed their teeth in friendly grins. After these opening conversational gambits, we got on like a house on fire. From our own goodly store we loaded them with packets of cigarettes, which they greedily and gratefully snatched from our hands, and searching their own almost empty pockets in vain for a return gift. One of our chaps got a bag of about 2 lb of rice, but they'd little else to offer in return. Not that we wanted anything, but chagrin and despair at not being able to carry out the Mohammedan first command — gifts and hospitality — was plainly written on their faces. Shrugged shoulders and empty palms displayed prominently indicated their inability, and their regret for their inability, to return our gifts in like manner. Our home-spun philosopher, George, did his best to console them. 'Well, Abdul, if y'ain't got fuck-all, y'can't give fuck-all, can yer?' I doubt if the exact meaning of George's words were translatable into pure Aramaic, but George's friendly smile made up for the deficiency.

'If only these heathen buggers could speak English, we could have a right little natter, couldn't we?' says our Sergeant, regretfully.

Our Officers meanwhile looked on, frowningly and disapprovingly, but helpless. Anyway, what do Officers matter when you've got a chance to have a friendly word with the blokes you've been trying so hard to kill these last few months.

For about an hour we fraternised—like an Old Pals' Reunion. We noted with some surprise, and with lots of compassion too, their own deficiencies in kit and clothing. Many of them had their feet wrapped in cloth wrappings in lieu of our stout army boots, and more than half of them didn't possess a greatcoat. Their Officers and NCOs, however, were by contrast, very well fitted out and clothed, but for their poor 'other ranks' it was a wretched state of affairs. 'And to think that a bloody crowd like this could fight so fierce,' one of our blokes wondered aloud.

I think it never occurred to us that they were defending their own soil. And whatever you say about Johnny Turk, he's a real stout fighter. I doubt if there's a better in the world, except of course, us!!!

The rain eventually began to slacken and just as surely our enforced intimacy began to slacken. As a last humorous gift one of our chaps dashed back to the trench we had just been forced to leave, and managed to find a small sack of tins of jam—our own beloved (!) 'Plum and Apple'—and depositing at the feet of one of the Turkish NCOs, this small sack of tins of jam, he remarked, "Ere y'are, Abdul, eat this bloody stuff and then y'll want to surrender quick,' and laughed outright as the Turks scuffled to grab a tin. 'Wait till they eats it,' he says, 'an' then Ah reckon they'll 'ate us worse than ever.'

Eventually the floods in our trenches subsided and we climbed back into them quiet and thoughtful. The episode of the last hour or two was fresh and startlingly new in our minds and nobody said much at first. The trench was still ankle deep in mud, but ankle deep mud was a minor discomfort in the pattern of our life. We retrieved what we could of such kit as had been washed about by the floods of water, laid out our groundsheets on the wet firing step and eventually shivered ourselves to sleep, crouched sitting up with our heavy water-logged greatcoats spread tent-wise over us. In a rare act of consideration and kindness, our CQMS sent us up an extra jar of rum to ease our clattering bones and quiet our chattering teeth. No guns spoke that night — it was like as if the artillery of both sides had conspired to say, 'Let's give the poor sods a rest tonight — it might give 'em time to ponder on what they've just seen about this lark they call The Brotherhood of Man.'

I can't even remember if sentries were posted that night, but I should imagine not. I guess we all felt, even from our Sergeant downwards, what's it matter if we are attacked and even killed — nothing could be worse than this. And I guess that went for the Turks, too.

Just before the arms of an icy-cold Morpheus enfolded us into arctic unconsciousness, we were consoled once again by our pet philosopher. George had taken off his boots and was busy pouring the water out of them on to the trench floor. There's one thing about our George, at times of stress he's at his best and most solemn.

'Ah allus think, y'know, lads, that there's allus some poor bugger worse off than thee thiself. Now take them poor bloody Turks, like. Why — some of them poor bastards ain't got no boots to pour water out of of.' This cryptic note of consolation left us puzzled — and grinning.

The front was quiet for the next two or three days and those in our immediate sector were invited to the rare treat of a share in a rice pudding. Well, so it was flatteringly designated — this mess of Turkish gift-rice, boiled in a two-gallon dixie and laced with a tin of condensed milk that one of us had nicked from the QMS stores. Still, if you shut your eyes, and wished back to the days when your Mother used to make 'em, and if you allowed for the inevitable difficulties of war-time culinary — all this and a lot of imagination, too — well, it was just like eating a real rice pudding. Well — nearly! Anyway, it was good enough to move George once more to the profound depths of philosophy.

The more I think of George the more I am convinced he was a direct descendant of Socrates or one of those old-time Greek philosophers who always had the answer for everything. But for an accident of fate, George might still be holding forth somewhere in Athens to a group of dialectically minded young men engaged in their national pastime of verbal hair-splitting. But for this same accident of fate, George might even have been calling for 'Another round of hemlock, landlord, for the boys and charge it up to me,' instead of handing round his water bottle half full of well diluted rum. Somewhere along the line George had got pushed off the conveyer belt and found himself 'piecing-up' in a Lancashire cotton mill, instead of being where he really belonged. The last grain of rice digested, the rum and water now exhausted, George is again moved to profundity.

'It's a funny thing, mates, when yer comes to think of it.'

'Wot is?'

'Well, this 'ere war. Now you take this rice-pudding, like. We wouldn't 'a bin eatin' it, but for that Turk wot give us the rice, would we know?'

'So?'

'Well, afore 'e give us this rice, and afore we'd met 'im we'd 'a killed 'im if we could.'

'Of course.'

'But 'e gives us this rice, and we didn't kill 'im. And we gives 'im cigarettes, and we're all friendly, like. Just like old pals, we was. And if we meets 'im again, wot does we do?'

'Well, wot, George?'

'Well, we kills 'im again, don't we?'

Someone reminds George that even a Turk can't be killed twice, but George brushes this off as frivolous and closes the conversation — 'First we tries to kill one another, then we're all friends and 'appy, like. Then we tries to kill one another again. It's a very funny thing, is war.'

The ponderous puzzle of man's greatest sport overwhelms him and he seeks his virtuous couch in slumber, leaving us thoughtful and uneasy. The night is young and the rest of us are disinclined for slumber yet awhile, and in youth's confident solution of age-old problems, we assess the merits and demerits of war — pontifically, arrogantly and illumined by the flash of youth's cocksure knowledge of things hitherto kept dark from the rest of mankind. One of our gang was a confirmed atheist, except perhaps just before a big offensive, when he'd seek spiritual consolation from some religious pre-battle service. 'Always believe in backing a horse both ways that's me,' he used to say — and grin.

'All wars are basically the result of religious intolerance,' he asserted. 'And what's the first principle of all religions? Well, I'll tell you — the first principle is to ram it down the other chap's throat, whether he wants it or not.' A student of history, he was well genned-up on his subject, and kept us enthralled with a racy account of man's activities in spreading the gospel of loving kindness. He touched lightly on the Holy Crusades against the Infidel Turk — 'like we're doing now but in another sort of way,' and led us through the infamies of the Inquisition, and its counter Protestant reprisals — 'all in the name of God and Eternal Love, mark you.' Strictly fair, he brought in the history of the Mohammedan intolerances, Sect against Sect, and against the Christians. 'And how did we get our colonies? Tell me that, if you can. How did we get 'em?'

'Well, Ah don't rightly know. Ah s'pose we conquered 'em or summat. 'Ow does ta reckon we got 'em, then?'

'Oh! lots of ways, ' he said, and chuckled. 'One way I've heard was to land the missionaries to tell the natives all about God's love and the Paradise to come, and while the natives' heads are bowed in prayer, up goes the British flag over 'em.'

Bill Hodgson, one of our Company Grenade throwing party, spat disgustedly. 'If all that tha sez is true, Ah'll never go to another Church service, that Ah won't.'

But our atheist grins again. 'Oh, yes you will, lad. We're having a grenade attack in a couple of days' time, so they tell me,' and Hodgson scowls and looks uneasy.

Our sergeant who had been lounging nearby in idle listening to the conversation, looks a bit doubtful at the way the conversation's drifting. Had it been some 50 years later, he'd have thought it to be bordering on Communistic-inspired talk (that's what we call it nowadays when we meet up with any unorthodox line of thought) — but at the time he probably thought it to be perhaps a shade disposed to lèse-majesté, and rather disloyal. 'Time we all turned in and got some shut-eye,' he says, 'you can finish yer talks about religion and paradises in the morning. But for me, mates, best bit o' paradise is a dry firing step for a bed, and to dream of wot Ah'll do to the Missus when Ah gets 'ome.'

Ay! The Sergeant's right. A dry firing step for a couch, and with your groundsheet pegged slantwise on the trench-wall, just above you, so that if it rains you're snug and dry below in your improvised tiny tent, and your blankets pulled tight over your head to get a nice warm fug up, and hear the bullets during the night whistling harmlessly over you, and all of you chaps who've ever done any trench warfare will know only too well that feeling of blissful unconsciousness that comes from a warm glow. Even your icy feet encased in your wet boots borrow some of the warmth stealing over you, and you feel so good—you wouldn't swap places with the King of England. There's no sensual pleasure so keen as to gradually feel your wet and ice-cold body thaw into drowsy warmth.

Just before you pass out to blissful unconsciousness you re-live the last 48 hours and the strange events of the alternating, unnatural, dutiful hate of the enemy, and the more agreeable aspect of mutual fraternisation like we'd just experienced. You get a bit wistful—wondering why we couldn't always be such good friends. If it was so marvellous in spite of all that drenching rain, it'd be even grander in fine weather, and you wonder why it's necessary to hate the enemy so much—decent, ordinary, little blokes like ourselves, too, they seemed to be.

As the sandman sprinkles his sleep-dust thicker still over your eyes, the vague figures and faces of those with whom we'd shaken hands, back-slapped, talked to, and exchanged our little gifts not 48 hours beforehand, become fainter and fainter, and the whole incident of the spontaneous demonstration of the real Brotherhood of Man fades into an unreal fantasy of the mind.

Then the problem gets too big for a boy-soldier to handle. Your trust is firmly anchored in the older men... the High-Ups, the Generals, the War Office—in fact, the whole bloody shower of those older and wiser men who have taught us 'tis right and British to hate and fight.

Ay! Ay! Ay! Maybe, boy soldier, that last little bit of enforced fraternisation has spoiled you forever from being a good soldier. Next time you draw a bead on some poor bloody Abdul, you'll be thinking 'that could be the same poor shivering bugger I gave some fags to once, that day we got washed out of our trenches,' and you'll probably make sure your shot misses.

And that's no good for war... as well as being so very un-British.

Damn all fraternisation—it spoils you from really enjoying a war properly. Some two years later, in a different Arm of the Service and in a different theatre of war deep in the Balkans, I'd spend my days in an Observation Balloon, working in close co-operation with the heavy artillery on the ground. But now, as a Commissioned Observer-Officer of the Royal Flying Corps, I was able to take a different view of the war. As a Tommy on Gallipoli, it was a worm's eye view. But now it is the lofty view from the air — the view of the bird of prey. The view of a tawdry, tinsel god swinging unconcernedly in his little balloon basket, decreeing Life or Death for some poor bastards down below. Up there, it's beautifully quiet — so still, in fact, you can hear your own heart-beats. Not a soul or a thing nearby. You're alone, so quiet, so high up — uplifted into a soundless universe. Occasionally the view of the earth below is momentarily obscured by the little clouds of white cotton-wool looking stuff that cluster just below the balloon basket, giving you the illusion that you could get out and walk on them. If you pick up the telephone you can have, if you want to, a buzzing crackling, field-service-telephone chat with the lesser and unworthy mortals below on Earth. But it's so quiet and serene up there, so near to Heaven, so cut off from the Earth, that you feel you could — again if you wanted to — 'reach out and touch the face of God.'

There's no wind, and the balloon basket merely rocks gently to and fro, as gently as a yacht rocks on the waters of a sheltered harbour, and you wish you could spend the next four hours of your duty up here, all on your ownsome, and not have to bother with the blokes down there on Earth. Just you, yourself, and the Big Fellah, even higher up still. There's lots of things you feel you'd like to sort out with Him, and where better than in a balloon basket — in a quiet sky — and half way up already to Heaven. Where He can hear you better, away from all the cluck and noise of the Earth and the quarrelsome insects that infest it.

The telephone crackles. It's the CO, Captain MacLean, warning me that an unidentified aircraft is around, and to be prepared to jump if necessary.

Damn! I hate jumping, although I've had to jump quite a few times. And I never could grow out of the shame-making habit of screaming my head off with fright as I dropped the first two or three hundred feet before the parachute opens automatically and when at last I feel the welcome tug on my harness and hear the welcome flap of the opening parachute as it opens — like an unfolding umbrella — above my head. Fortunately, no-one could possibly have heard my screams, except perhaps the Big Fellah, and I guess He got many a quiet chuckle. 'Does 'em good to get a good scare now and again, the cocky bastards,' He mutters. But once the parachute opens, well — then it's worth the first fright of the drop. There's the long slow slide to Earth, blissfully quiet, blissfully peaceful.

You've got to actually experience this slow peaceful slide to Earth at the end of a parachute to know the feeling. Ask any parachutist.

So I gather up the maps and field-glasses and keep my fingers crossed. The 'phone crackles again and I get ready to climb over the basket side. Then I hear MacLean's chuckle. 'That put the wind up you, old boy, didn't it? But relax — it's one of ours. How's the visibility?'

'Lousy. Little banks of clouds just under the basket. Can't see damn-all.'

'It'll clear. It'll clear very soon. Go back to sleep again.' And again MacLean's evil little chuckle.

Some folks have a queer idea of humour, I ruminate peevishly.

But the little white bank of clouds has now disappeared and the view of the earth is clear. The blasted telephone buzzes and the Siege Battery says it's now ready for the map co-ordinates and then essays its first ranging shot on the target they can't see. Methodically, carefully, you correct their fall of shot — 'lengthen 500 yards, at 11 o'clock, etc.' This clock code's dead simple for directing fall of shot and especially easy from an Observation Balloon, and soon you're rewarded by a flash of flame in the target area and through your field glasses you can see the enemy battery and its ammunition has gone up in flames and splinters. You can't hear a thing, but you can see it, and you — you tawdry little tin god — vaingloriously you've borrowed some of your Bigger Brother's attributes and decreed who shall die below. You feel loftily superior to other earthbound mortals, and when back on the ground again you receive even your own CO's congratulations — 'Bloody good show, chaps' — you receive it loftily and off-handedly.

Later in the Mess that night you get a nagging reminder of your Gallipoli fraternisation. Even in this different theatre of war, this nagging reminder spoils the evening for you and makes you wonder how many little Bulgarian brats you've made fatherless today, or how many widows will now have to pad the sleazy districts of the dim lamp lights later on, just to get enough to eat for their kids. At odd moments now you don't feel so bloody clever. Fortunately, booze is cheap in the Mess and we're a boozy crowd. It makes it easier to forget.

'Why so glum old boy,' the CO chaffs me. 'You should be on top of the world. Battery Commander says it's the best shoot they've had in weeks. So cheer up and stop looking like a wet week.'

'Oh! — I dunno, Sir. I was just thinking.'

'Deadly stuff, thinking, old boy — deadly stuff. Have another Scotch.'

So to hell with all fraternisation, I say. It doesn't make a better warrior. Whether it makes for a better human being is a thing we've got to decide for ourselves.

As we get older and look back, and begin to wriggle a bit uncomfortably when we think of all the suffering which we, individually, may have caused in the sacred name of duty — well — somehow it doesn't seem so clever. How many men have been crucified on duty. This sacred cow of Duty. I dunno. Much of man's noblest exploits bear the familiar tag of Duty. And much of his less worthy exploits, too.

'Nothing personal in this, yer Majesty,' the headsman apologises to King Charles I, before wielding the axe. 'Just in the line of me dooty, as yer might say.'

'Help me nail this Bloke to the Cross, Flavius,' says the Roman soldier. 'Not that I want to specially; He seems harmless enough, poor old chap. But duty's duty. And that damned centurion is watching us, too. So pass me some more nails, will you?'

This sacred cow of Duty. Maybe it's all part of the Divine Plan. I wish I knew.

> Before beginning, and without an end,
> As space eternal and as surety sure
> Is fixed a Power Divine which moves for good,
> Only its laws endure.
>
> It slayeth and it saveth, no-wise moved
> Except unto the working out of doom;
> Its threads are Love and Life; and Death and Pain
> The shuttles of its loom
>
> It maketh and unmaketh, mending all;
> What it hath wrought is better than what had been,
> Slow grows the splendid pattern that it weaves
> Its wistful hands between.

But I still wish I knew... and for sure.

CHAPTER 27

An Innocent Abroad

I don't know whether it is peculiar to me, but whereas I can remember vividly periods when the sun shines, my memory gets clouded over when I try to recall long periods of intense cold. It's like as if memory itself is put into cold storage when the temperature drops way below 32° Fahrenheit. I wonder if it's the same with other people. Can you remember vividly and in detail any blanketing bitter winter, as easily as the long hot summers? Intense cold not only freezes the body. It seems — and probably mercifully — to freeze the mind and memory, too. In his epic account of Gallipoli, Alan Moorehead, quoting the words of some war correspondent on Gallipoli at the time, notes: 'Triggers were jammed and rifles refused to fire. Sentries were found in the morning still standing, their rifles in their hands, but they were frozen to death.' And of others, 'they could neither hear nor speak, but stared about them like bewildered bullocks.'

But they were only some of the troubles. For me, winter was a blanketing period of time when it's so bloody cold that life is just an endless stamping of feet on the trench floor to stop your legs freezing solid, the cupping of the fingers tight close around your burning cigarette tip, just to thaw them into a semblance of life. The grateful dipping of your tin mug into the dixie of tea simmering on the little fire, and while you lay your mug down on the firing step for a split second or two to reach for a fag, the tea's frozen solid before you've even time to light up. Your mate speaks to you but before you can hear him you've got to unwrap that bit of cloth from around your ears, or push the woollen

CHAPTER 27

Balaclava helmet back from your head so you can hear what he says, and then watch him repeat the performance if he wants to hear your reply. Most times you're too numb, too cold to want to either speak or listen — too cold to either want to live or die. Just frozen immobility, the frozen lethal Death-in-Life. And come nightfall your bed of the firing step sparkles with thick frost. You loosen your bootlaces, lay your aching bones down, pull your blanket tight over your body and face to try to get a warm fug up, just enough to thaw out your ice-cold limbs. Your mate on two-hours sentry-go piles his own blankets tight round you — so tight in fact that you're wrapped up like an Egyptian mummy, the risk of suffocation being far the lesser evil. In time, and if you're lucky, you'll feel the grateful thaw steal down your legs and know for a fleeting short time the exquisite ecstasy of warmth — the bliss of restored circulation. No king on his throne is half so happy as you. Because you're warm again, warm, and drifting into the sublime forgetfulness of sleep. All too soon your mate gives you a shake. It's your turn now for sentry-go so you do the same for him — wrap him up tight, stifle your envy, then stick your frozen face to gaze over the trench top at the barbed wire of No-Man's-Land. It looks extra peaceful tonight — all that thick frost on the barbed wire, and stretching all along the line in front of you — like the Christmas decorations of a big hall.

Christ! but it's cold. The keen North wind from the direction of the Turkish lines makes you duck your head for a few seconds below the level of the trench top for shelter. The sound of the approaching Duty Officer and Sergeant reminds you you're supposed to be keeping a keen look-out. You brave the icy wind again and lean properly over the trench top, rifle in your frozen fingers pointed enemy-wise, and wondering privately if Johnny Turk is doing the same damn-fool thing, this bitter night. The Officer appears, his chattering teeth mouthing the usual enquiry, 'Anything to report?' If your brain wasn't so completely frozen up you might be able to think of something crisp in reply — something like 'Thousands of silly bastards like us, Turks and British, getting a hell of a kick standing here, slowly freezing to death,' but instead you make the usual reply, 'All quiet, Sir,' and he disappears around the next traverse of the trench, and you duck gratefully below trench-top level again to shelter from the icy blast.

Christ! but it is cold — colder than ever. Your fellow sentry says something, and after about three tries you manage to hear what he is saying — something about 'Shouldn't be surprised, mate, if there's snow before morning. Eee, but wot wouldn't Ah give to get on that bloody hospital ship,' and he essays a few minutes 'one-armed sentry-go.'

But it's no use, soldier, you never get 'em where you want 'em. No use — this 'one armed sentry-go' lark — holding high above the trench your arm, waiting for the grateful enemy bullet to plough into your arm so you can get on that hospital ship lying there so temptingly off-shore, all brilliant lights and its Geneva Cross blazing on its funnel. Luck doesn't work that way, soldier. When you want a nice cushy Blighty one in the arm, like as not you'll get it where you least want it. And you know where that is, don't you, soldier. And eventually your fellow sentry gets tired of playing this 'one-armed sentry' game. 'Can't shoot for toffee, these bloody Turks,' he says despondently, and lowers his arm despairingly; we chip him unmercifully but take the sting out of it by imitating him, till a dozen pairs of arms are held high up in the freezing night air, and we mock the enemy's shooting in caustic Lancastrian humour.

"Ere y'are Abdul, three shies a penny.'

'Hit my bleeding arm and win a cokernut,' we bawl out across the quiet of No-Man's-Land.

But the elusive bullets fly by us unheedingly.

By next morning the hospital ship has pulled out for Alexandria with its usual load of sick and wounded. She'll be back within three or four days for another load of this endless stream of patients lying in its hospital bunks, some too far gone to care at all — some hugging their secret joy and giving private thanks to the gods for the blessing of a cushy and none too serious wound, this blessing in disguise that is getting them away at last from this frozen hell ashore. The hospital ship is warm. And clean. And there's clean sheets and good food to make your small wound bearable, and that is a whole lot better than being frozen to near insensibility, half-starved on unspeakable rations, and spending your days and nights in a foul trench — too cold sometimes to make a trip to the bog-hole, and if you do make the trip, it's nearly half an hour's work to get your braces unhooked from your trousers, 'cos your fingers are frozen nearly solid. So cold that you wonder if you'd ever see the next day through, and not caring too much whether you do or not.

'Just like being at 'ome on that ship, mate,' one of the chaps told me who had had a two months trip away from us in an Alexandria Military Hospital. 'You can get a bath, too, on that ship — real hot water, and all the soap you want. You can get a bath any time you want — bath all day if you want to. And clean towels, too. Ay! it's grand on that ship, mate. And when you gets to Alex they're all over you — the English people there think you're a bloody 'ero, instead of an ordinary louse-bound soldier wot doesn't give a monkey's f— which way the war's going.'

'But comin' back to this bloody Peninsula after a two months' hospital stay — cor! it's hell, mate. You feel yer can't face it again. You do an' all. It seems like a dream — that trip to Alex. mate. Just like a bloody dream.'

That bitter winter sure did demoralise us, coming as it did after all these fruitless months of wasted endeavour, wasted fortitude, and all too often wasted gallantry. We couldn't have been at a lower ebb.

The snow held off for a day or two. The night before it fell a young and happy soldier was returning from the beach to his own sector of the front line. It was a cold crisp frosty night. The stars in the crystal-clear sky seemed twice as big. And the soldier was tolerably happy. He must have been, in spite of the intense cold. He was whistling cheerfully as he shouldered his precious burden back towards his Company. The exercise, the scrambling over the rough ground, climbing steep terrain of the sides of the gully — all combined to make him feel warm and good. It had been his turn that night to make the trip to the beach for the bi-weekly rum ration now carried so carefully in that gallon stone-jar on his shoulder. Life it seemed, was almost bearable again. His body glowed from the exercise and effort, and the upsurge and optimism of youth once again rose triumphant. He called out a cheery goodnight to the two middle-aged soldiers of the Signals Section, squatting in front of their dug-out on the hillside and toasting their hands and feet over a small fire. A mess tin of coffee bubbled quietly at the side of the fire and the fragrance of the coffee rose like a song on the cold quiet air of the night.

"Ere, soldier — like a cup o' cawfee?' and the boy stops. Gosh! but life is good. A beautiful clear night like this, two decent old chaps round a little camp fire and above all the friendliness of every thing. So he stops and gratefully sips his coffee, chats and smokes. They can't do enough for him, these two 'old sweats.' One of them glances slyly at the stone gallon jar and wonders audibly 'how a drop o' that stuff, mate, would go down with the cawfee.' Regretfully the boy explains he didn't dare open the jar. 'It's a sealed jar, see, chaps, and I've got to deliver it to the Company Commander tonight for the lads' rum ration.' They agree solemnly, and sigh. Then one of them, and in reminiscent vein, starts telling about a pal of his, 'carrying one of these 'ere jars, just like you are a doin' mate, 'appened to stumble, like, and the seal got broke. Not the jar, mind you — you'd want a sledge hammer to break them bleedin' jars. But the seals... they break that easy, ee they do that.'

The boy is uncomfortable. The temptation is great and the wrestling match goes on inside him. He's one of those ordinary honest Joes with a strong sense of duty. But he's also cursed with a strong 'noblesse oblige' disposition. And when you get a boy soldier like that and in these peculiar circumstances, you can lay the odds that 'noblesse oblige' will pip Honesty and Duty by a short head. So the seal 'gets broke, accidental like' and a generous dollop is measured out all round and unfortunately including the boy, and the deficiency in the jar is replaced by an equal amount of water. All would have been well had it stopped at that, but one hundred percent-proof rum contains within itself more hidden snares than a rabbit-poacher's wood on a moonless night. So the now unsealed stone jar is unscrewed again and another round is measured. And yet another. And another. And that's the last the boy soldier remembers.

When I wakened next morning, I could scarcely move for the weight of cover on me. With any slight movement of my head the pain was excruciating and when the mists and fog had cleared from my brain, I realised that while I slept a heavy fall of snow had covered me. With difficulty I managed to free one arm, and after a struggle — and a long and painful struggle it was, too — managed to brush the weight of snow off my legs and body, and stand up. Every limb was racked by the Devil's own torture-rack; fortunately, my hot breath had melted most of the snow on my face and head and saved me from suffocation during the night. Thoroughly alarmed at my condition I managed to contact the Medical Officer in his little First Aid dugout on the hillside

CHAPTER 27

and told him I thought I must be suffering from acute rheumatism or something, and that I could hardly move. Maybe he was the unsympathetic type. Or maybe he was over-fastidious, for my rum fumes made him turn his head away quickly. I got the sick report to hand in to my Company Commander. I thought it a bit unkind, this sick report, marked 'Duty.' Duty on a sick report is about as bad as you could get. It means anything, but mostly it means 'deserving of no sympathy or medical care all — deserving of whatever punishment is seen fit.' So it was with some trepidation I handed it in to the Company Commander. He would have been justified in handing out to me the most severe punishment possible — maybe Field Punishment No. 1. Field Punishment No. 1 has varied throughout the ages in the British Army. Legend has it that at one time not very far back the favourite form of punishment was to strap the delinquent, standing, to a Field Gun, binding him hand and foot, with his back to the gun, lashed to the gun-wheel, and leaving him standing there bareheaded for about 4 hours to stew under a hot desert sun, or in the freezing cold. But the Company Commander eyed me sadly, and I told him the truth of what had happened. 'And where's the remains of the rum now, lad?' he asked. I told him I didn't know — lost — or probably pinched by those two 'old sweats' who'd been my accursed hosts the night before. He was a good old boy, our Coy. Commander — been in the ranks himself once, before he got commissioned, and knew all the pitfalls of service life.

'There's no punishment I'm going to give you. I wish there was — it'd be easier for you. Then we could call it a day,' he said. 'Unlucky for you though, you'll make your own punishment. I'll leave it to you to explain to your mates why they didn't get their rum ration last night — that is, if you dare tell 'em.'

They should have clobbered me, my mates. I deserved it. I knew all too well what it's like to be anxiously awaiting the life-giving tot of rum on a bitter freezing night and with no cover from the ferocious cold but the sky. They should have clobbered me and called it a day, instead of sending me to Coventry. And if the real city of Coventry is as cold, as bleak and as hatefully unfriendly as the Coventry I was sent to, then I never want to see it. Chaps looked away when I spoke to them, spat on the ground if I happened to pass — y'know — all the non-violence tactics that hurt worse than any clouting. Not that I blame them one bit. Had it been me, cheated out of my rum ration that night, I guess I'd have wanted to poison the cheater. But I'm the vicious type.

For about four days they cut me dead. Then I got my usual bumper parcel of Canadian pipe-tobacco, and dished it out to the pipe-smokers and the ice melted. By some alchemy, these older chaps, the pipe-smokers, even had me portrayed to the younger boys as the innocent victim of a dastardly outrage, and warm sympathy suddenly replaced the hitherto frozen mitt. Which just shows how illogical our fellow men can be. But if they were logical and sensible, damned if I'd like 'em half so much. It's the unpredictable, the illogical element in mankind that makes 'em so interesting, don't you think?

CHAPTER 28

God Rest Ye Merry, Gentlemen

A frozen blank blacks out the memory. Occasionally one remembers some of the teeth-chattering days, the sublime misery of being cooped up in a trench with 20 degrees of frost as a companion, a keen wind that searches your very bones when you nip to the bog-hole away from the protection of the trench walls. It's too cold to take off your shirt for the daily de-lousing act, so you scratch and scratch and scratch. Scratch until your whole body is a mass of pestilential small festering sores from your own foul and dirty finger nails.

Then a quick thaw for a few days, heavy drenching rain, and your greatcoat and uniform is a sodden heavy slime. A pool of water to sleep in. Even the frost is better than this, you think. Then the frost comes again to ask you if you are still of the same mind. Your sodden wet greatcoat now hangs about you as a frozen solid wall and you feel like a knight of old encased in steel armour. There must be an end to this some time, somewhere, but you're even too cold to think more than a few minutes ahead. Heaven, it seems, is now contained for a few minutes in the small and now daily issue of rum — that blissful few minutes of false warmth that alcohol gratefully provides. There are no teetotallers left now in our Company — it's either drinking the daily tot or going mad. A phenomenon of this weather is that no one ever catches a cold, and to this day I've never been able to understand why. Occasionally one of the blokes collapses completely and there is a rush of volunteers to help him down to the beach field-hospital. This rush to aid is not, I fear, entirely altruistic, for when you get him down to that little hospital on the beach you can get warm again for a few minutes in their little canvas marquee that serves as a hospital — thaw out your frozen fingers over one of the little paraffin heaters they use to heat the marquee. Maybe they'll even give you a cup of hot coffee, or you may be able to 'nick' a tin of condensed milk from the medical stores there. Anyway, the walk is good exercise and gets your circulation going again, and you don't hurry back to the icy chamber of your trench. 'If it wurn't for that bloody sea, I'd put miles between me and this bloody place,' says one, and lingers longingly on the possibilities of desertion. 'I'd walk for days and nights, if there wur owt to walk to, that I would.' But you're trapped. You feel trapped — trapped like an army of rats waiting for the pest officer to come along and exterminate you. The days pass in unending arctic procession.

Then all of a sudden and before you're even aware of it, Christmas Day looms up.[26] On the morning of the sacred 25th, you waken to find the whole war-scarred countryside has been blanketed overnight by a heavy fall of snow. From our high vantage point, we gaze down across the main gully. The ragged terrain has been smoothed off considerably by the soft thick blanket of snow and the snow sparkles in the morning sunlight. It's beautiful to look at and you feel a lifting of the heart as you gaze over the countryside. A stainless white blanket — hiding nearly all the ravages of war, stainless — except for the pink-splashed snow near your trench from the blood of the wounded as they had staggered by after last night's hand-grenade attack. Some super-brain, back in his snug dugout on the beach, had been doing his homework and reasoned that, being Christmas the Turks would be the last to expect a Christian grenade attack on Christmas Eve. Unfortunately for the grenade throwers, the Turk had been doing his own homework too, and had reasoned equally that being Christmas Eve it would be a good time to take advantage of Christmas relaxing. Result — a ghastly head-on clash.

Down below in the gully, the apotheosis of incongruity. Incredibly, on this Devil's own playground, someone had unearthed the Battalion's band instruments from their temporary storage and the bandsmen were

---

26 See Appendix III, Notes on events and places mentioned in the text.

gathered down there in the gully, their polished instruments winking and flashing in the morning's bright sun, and the strains of Christmas carols were wafted on the still morning air.

One of our chaps had a wonderful tenor voice. Back home he used to earn a few extra shillings at week-ends, singing in pubs, and he accompanied the music of the band in mellow and heart-searing voice. Accompanying the distant strains of the band he even sang, right through, all the verses of 'Once in Royal David's City' while the rest of us stood around in silence, staring into the void of space. Hushed and over-awed, we didn't dare meet the other's eyes. The magic of music. Panoramas of other earlier and happier Christmases rolled before us in a scenic flood and one of the younger boys sobbed unashamedly.

And me! Well, do you think I can ever hear this carol again without being transported to that particular Christmas Day on Gallipoli. I'm hearing again this chap of ours with that marvellous tenor voice accompanying the band's soft music that floats, unreally, from down there in the gully. Ever since that particular Christmas Day, all the beauty of all the world is, for me, contained in that particular Christmas carol.

But the music and the carols ceased. In the distance we could see the band moving off to another part of the gully, there to regale another lot of listeners with the spirit of Christmas. We sat around, quiet, for a bit — the blokes staring down glumly at their boots, scraping bits of mud off them, or twiddling with the tapes of their puttees — an awkward embarrassing spate of silence, every man deep-wrapped in his own private musings. It was left to Bill Howarth,[27] one of our quieter types, to break the silence. In a low voice he asked, 'Sarge, does ta believe there's a God?'

Startled, the Sergeant regarded the questioner curiously. 'That's a bloody funny question to be askin', Bill. Why, doesn't ta believe in Him thiself, then?'

'Ah'm askin' thee, Sarge,' said Bill pointedly.

'But why ask me? Ask t'other blokes. Why pick on me, mate?'

But Bill with an impudent grin said, 'Well, tha's a Sergeant. Tha ought to know better'n me. Ah'm nobbut a bloody private.'

---

27 Probably Pte 8593 W. Howarth. See Appendix II, Notes on individuals mentioned in the text.

'Ah know one thing, Bill—tha's a cheeky young bugger. But wot makes thee ask, lad?'

'Oh! Ah dunno, Sarge. Being Christmas Day and all that. Just askin' like. Does ta believe in God?'

'Well, Ah suppose there must be one somewheers. That's wot they all say, anyhow.'

'But does ta thiself, Sarge, believe in God?'

The Sergeant stood up, fishing in his trouser pockets for a fag, while his gaze roamed over the parapet at the war-scarred terrain—the tops of the wooden barbed-wire supports protruding over the top of the snow—the distant view of the untidy improvised cemetery—one of the many—its numerous crude little wooden crosses now almost invisible under the fall of snow—the ever present view of the hospital ship lying just offshore, waiting till she gets her full complement of sick and wounded before making her periodic trip to the Military Hospital at Alexandria. Maybe, too, his eyes saw lots of other things—otherwise invisible things except to soldiers—lads who'd 'handed in their checks' in bewildered agony—lads who still lived all around him, in uncomprehending and uncomplaining misery and despair, stoically sticking it out—drinking deep of the corrupting chalice brimful of the baleful brew of what we complacently label 'man's inhumanity to man.' Maybe he saw lots of other things that go to make up the sport of war. His expression was hard and bitter, his lips twisted and tight.

'Look out theer,' he said harshly, 'and then tha asks me if Ah believe in God.' He stopped short and none of us dared break the pregnant silence. We waited, expectant-like, for him to continue. Rarely had we seen our Sergeant so moved. Then he looked round at us, his eyes roving us all appreciatively. His face lost its bitterness and his lips loosened in a little smile. An unlettered man of little learning, our Platoon Sergeant, but maybe even he too saw the invisible aeon-old battle still raging between the primordial cruelty of man and the slow emerging forces of love. In softer tones, and picking his epithets deliberately to offset the gruff affection of his voice he continued, 'And then Ah looks round at all you scruffy lot of buggers wot's bin me mates ever sin we left England—well, anyway, them of you as is still left—and when Ah thinks of all we've bin through together, thee and me, lads, and the way we've bin to one another, like. Like brothers we was, all the time and every one of you buggers doin' the best 'e could. And then maybe, you asks me again, like, do Ah believe in God... so mebbe, then, Ah thinks a bit different, like. An' mebbe Ah says "Yes".'

Nobody spoke. This unexpected 'unsolicited testimonial' left us speechless with surprise and gratification. Then he turned his back on us abruptly. Standing on the firing step, head and shoulders above the top of the trench, he gazed at the long stretch of terrain — from the ragged coastline, off which lay tantalisingly the coveted hospital ship. His eyes roamed the white snow-covered Desolation of the Damned we knew as No-Man's-Land, and finally coming to rest on the untidy cemetery where slept so many of our boyhood companions — boys who would never come to know the indignity of old age. He stared for some time — as if hypnotised — on this cemetery, this travesty of a Temple of the Dead, and where so many of our fellow townsmen lay.

Two long rows of little crudely fashioned wooden crosses. There were some gaps in these two rows of crosses — result of the odd stray shells falling there accidentally. In the moonlight, these little crosses used to shine like the gleaming white teeth of Death itself, and with the gaps looking as if the Supreme Jester had had some of his own teeth knocked out in the burly-burly of war. We used to watch these odd shells that fell there accidentally at times — see some of the crosses and some of the bits and pieces of the chaps lying there, tossed high into the air. With many a macabre jest, we'd speculate how these blokes would be able to sort out their own bits of legs and arms at the Last Day. At such times and with something of a shock you felt within yourself the faint stirrings into life of a fiendish delight in the savagery of it all — this final obliteration — as the sacred remains of those who had sought final refuge and peace under a Christian cross are tossed mockingly, high up into the air. And then you realise — again with something of a shock — that not a lot divides you yourself from the Palaeolithic age.

Under the weight of the snow some of the little crosses leaned sideways apathetically. Many of them were only vaguely discernible under the cover of the drifting snow. The names on these crosses, once carefully inscribed in indelible pencil, would by now be completely indecipherable from the rains and the weather. These erstwhile companions of ours, their exact identities now completely and irretrievably untraceable — they lay there in the cosy anonymity of the dead. But somehow the very ugliness, the cold horror of this last resting place had a beauty all of its own. These graves, unadorned by marble headstone, 'storied urn or animated bust' — they had a powerful stark simplicity — a close comradeship of rudely fashioned wooden crosses, crowding together in the snow as if for mutual warmth — it all had a powerful primitiveness that bit right deep down into your heart. Almost you could imagine them — these mates of ours — under that snow, chatting and joking, swapping reminiscences, grumbling at the cold, complaining bitterly of the lack of blankets, and waiting for the rum ration to be dished out.

I reckon our Sergeant must have been staring at that crazy little cemetery for a good three minutes, saying 'nowt,' just looking. God alone knows what else he saw as he stood there, just staring. Maybe he even saw the feet of the great god Mars himself, gloatingly padding around in the entrails and excrement of mankind that littered the floor of His Own Private and Abominable Slaughterhouse.

Like a figure cast in stone, our Sergeant stood there immobile for a good three minutes. Talk about hearing a pin drop! I've heard the quiet of the desert itself, but I never heard anything so quiet as this — what with the snow — well — y'know how a heavy fall of snow makes the whole world go dead quiet and hushed — like as if a thick noiseless carpet had been laid over the earth. And us — the grubby spectators of our own Sergeant, himself caught up in contemplation of a drama of history in which he was just another part-time actor. Us — his grubby unkempt spectators, with our long matted hair and unwashed dirty faces — strangers now to soap and water for the last three or four weeks — hushed spectators awaiting the finale of the Sergeant's impromptu lines.

Maybe he was recalling the highlights of this epic motion picture in which we small-time ham actors had played our parts, or filled in the crowd scenes — an epic motion picture — a saga of high hope and endeavour, historically unique in its colour, its drama and its tragedy. First flashed on the world's screen in the 'glorious technicolour' of a Mediterranean spring, but now reaching its dramatic conclusion in the drab black-and-white winter of despair and defeat. In this living cinema there'll be no rush for the exits when the national anthem plays. We'll crowd out in orderly fashion. We'll leave this dramatic spectacle of history — not to the strains of the national anthem, but on the lilt of a bawdy soldier song.

This live film rolls out its reels of history till the last bit of celluloid is done, and the projector light dimmed. Soon — very soon now — this live auditorium will be as empty and as ghostly as only any auditorium can be when the show's over.

While we watched our Sergeant, none of us said anything. There wasn't a bloody thing you could say that would sound even the least bit appropriate during these tense few moments so we took refuge in silence. But the emotion's so thick, you could cut it with a knife.

Finally, the Sergeant turned round and faced us, rubbing his eyes with the back of his hand — muttering that the light from the snow 'makes yer bloody eyes watter, don't it?' and sat down alongside us again. In a low voice, like a man talking to himself, he said, 'An' then mebbe when Ah thinks of all them poor bastards lying out theer — good Lancashire lads wot's never done no harm to nobody... an' then you asks me again, like, do Ah believe in God...'

With a sigh that came up right from the bottom of his boots he answered his own question. 'Do Ah believe in God? Do Ah believe in God? Honest, mates, Ah don't-bloody-know. Ah just don't know. And now,' he said brusquely, 'wot about a tot o' rum?'

And so we shelved the deep metaphysical question for good, consigning it for answer to the winds, to the vasty deep, to the birds and the bees, to all the 'things bright and beautiful, and all the creatures great and small' who probably know more about it instinctively than all the platitudinous parsons, the ponderous philosophers and the profound professors of metaphysics. And who would certainly know more about it than our own half-starved, half-frozen, lice-ridden soldiery with no reserves of philosophy to fall back upon other than their own golden bond of comradeship and all that that implies to the soldier.

Then we came to life again and celebrated the birth of the Prince of Peace in a double issue of rum. We cleared our throats, and in cracked voices wished one another a Merry Christmas, quite unconscious of the irony. Loudly attributing our wet eyes to the effect of the cold, we too asserted our manhood by complaining bitterly that the light from the snow 'made yer bloody eyes watter.' Lacking handkerchiefs, we wiped the water from our eyes on our tunic sleeves.

Says George Parker, longingly, 'Fust thing Ah'll get me when Ah gets to Cairo will be to buy meself a big plate of fried eggs, stacks and stacks of good bread-and-butter, and a pot of tea — in a pot so bloody big you'd be able to swim in it. Wot does ta think o' that for a start, Sarge?'

The Sergeant regarded the questioner quizzically, then burst out laughing.

'Knowing thee as Ah do, mate,' he said, 'that'll be the *second* thing tha'll be gettin' thiself.'

Then we all joined in the general laugh, for Parker's prowess with the bints of Cairo was legendary. We discussed hopefully current rumours of the impending evacuation of this heartbreak place of frustration and despair, and after the manner of our kind, gloated obscenely at renewing acquaintance with the bints of Cairo and Alexandria.

CHAPTER 29

Run, Rabbit, Run

So Christmas came and went. The oven-browned turkey, the pudding alight with brandy, the beer, the whisky, the nuts and wine, the lighted Christmas tree and presents — in fact all the things we didn't have this Christmas — we didn't even miss 'em. We had a far better Christmas present looming up. And all dead official this time — no mere rumour now.

In a few days we were to evacuate this graveyard of British and Colonial endeavour, and it was to Egypt we were to return. Egypt and its hot sun. Heaven itself seemed a poor substitute for the sun of Egypt.

There was a very heavy Turkish attack a few days before the evacuation — it was almost as if the Turks had got wind of our plans. It was during this attack that our Machine Gun Sergeant and Corporal were killed, as I mentioned earlier.

An incident occurred, too, with our Battalion the day before the evacuation — something that gave us the opportunity to unload quite a lot of our pent-up bitterness. That was when the Quartermaster's stores section came round to our trenches, laden with brand new oilskins and sou'westers. Each of these would have been worth a king's ransom if we'd had 'em earlier, but it was said they'd been lying there, in bales, unissued, at stores HQ on the beach for some two or three months. But now faced with our evacuation and lumbered with all these waterproofs, some decided they should be issued individually to us rather than let Johnny Turk get them.

So we told the QM and his staff what to do with 'em and where to put 'em. Not so much because we had quite enough to carry already of our own equipment, but the cruel irony of it all savaged us worse than the slash of a knife. We'd have damned near wept for joy if we could have had 'em some 2 or 3 months ago. But somebody had sat on their fat arses in a snug billet on the beach, finding it too much trouble to issue them — either that — or they had been awaiting orders before they could properly make the issue. But now they were an embarrassment to the QM's staff so they decided to burden us with them — and just before returning back to the hot sun of our beloved Egypt.

The front-line soldier is all too often short of things but he never lacks for adjectives, and these we distributed freely to an outraged QMS and his staff. Threats of punishment for 'Disobeying an order' — 'Refusing to sign as having received an issue of oilskins' — 'Using abusive language to an NCO' — and even 'Mutiny' — we brushed all these aside like flies and laughingly told the QM's staff where to stuff their f— oilskins. It would have been a physical impossibility for them to have complied with our imaginative and colourful suggestions but we got great joy in telling them. And for good measure we gave the QM's staff our absolutely unsolicited testimonials of blokes who could see their own mates drowned like rats in a sewer, while they were too f— lazy to have issued these oilskins at the proper time. This unusual and very unmilitary outburst on our part restored our good humour tremendously, and long after a furious QM had departed in high dudgeon, breathing fire and slaughter, and leaving his piles of unissued oilskins in the trench, we laughed our heads off. I reckon it did us good, that incident — we got the sort of relief like you get when a painful boil is lanced. And many a grateful Turk must have blessed our QM's staff after we had sailed away. Someone suggested putting a booby trap under one of the piles of oilskins, but nothing ever came of it. By now we were past all bitterness. All we wanted was to get on those boats and away.

In the first week of January, ranks began to thin out, blokes departing silently to appointed rendez-vous on the beach. I never actually knew what happened or how it was organized, but chaps would be there with you one day in the trenches and the next day they weren't there. Came the last night and a call for volunteers — just a few — to man the sectors of our front line trench until the last possible minute while the remaining body of troops got away and to give the Turks a deceptive semblance of normal occupation and life — keeping up our usual nightly intermittent rifle fire.

I volunteered.

But if you've got any impressions of undue valour on my part, you couldn't be more wrong. What swayed me at the finish was the present of one of those Ordnance-issue wrist watches. The half-a-dozen in our Company who stayed behind were to be spread out — one to each half-a-dozen traverses or so — and to stay there with synchronised watches until it was the hour to cut and run for the beach and catch the last lighter pulling off the shore. 'And anyone 'as volunteers can keep the watch after,' our Platoon Sergeant told us.

I'd never owned a wrist-watch in my life, and these Ordnance 'Officer-issue-only' watches were wonderful jobs, stoutly made silver watches, accurate and good looking. No redskin savage was ever enticed so easily with a string of glass beads as I was with this watch.

The plan was to lay umpteen of the many spare rifles we had — lay them fully loaded — 10 in the magazine and one up the spout — spread them out along the trench top and the blokes staying behind had to run up and down the trench, firing them intermittently and haphazardly. The hope was that the Turks would be lulled into the feeling that the trenches were fully and normally occupied. And it worked, too. Unbelievably, it worked perfectly.

I've read somewhere that all sorts of ingenious devices were improvised to fire these spare rifles — long after the trenches were vacated — such as a tin attached by a string to the rifle trigger, and by a device allowing water to drip into the tin, the weight of the tin eventually triggering off the rifle. With due respect and realising that any soldier only sees what happens in his own stamping ground, I beg leave to be sceptical about this device. Maybe the originators of this story had been fans of Heath Robinson. By the time we'd have got these ingenious devices fixed and tested, anything could happen. In any case, there's always a lot of cock-and-bull stories floating about regarding times like these, and most of 'em invented by blokes who'd never heard a shot fired in anger.

Our plan was simple, practical, and called for no inventive skill. Zero hour, 'cut-and-run' hour, for me and my mates was 11 p.m. by our prized and coveted watches, after which we were to hare it for the beach. It was a damp night and the job of running up and down the trench firing these rifles made me sweat a bit. Further along our line I could hear the comforting sounds of other rifles being fired, so I knew my mates were on the job, too. By about 10.45 p.m. there was an ominous silence on either side of me, making me wonder if my watch had stopped, but it was evidently OK. Mild panic sets in and I explore the set of traverses on either side of me. Not a bloody soul in sight. It was obvious my mates had anticipated zero-hour and, instead of waiting for one another, had hot-footed it for the beach individually. Not that I blamed them. I was often tempted to do it myself these last two hours. And now real panic sets in and I followed their example.

It was an eerie experience. I guess the first man who lands on the moon will feel like I felt that night. The loneliest man in all the world. And the most frightened. As I tore panic-stricken along the lines of trenches leading towards the gully and thence to the beach, it was like as if I was on a dead world.

> So lonely 'twas, that God himself
> Scarce there seemed to be.

I know now what the poet meant when he wrote those lines.

The tortuous, twisting and now empty trenches, normally difficult enough to negotiate even when occupied, were now more than ever harder to struggle through in my panic urge for speed, and every few yards obstacles of discarded kit hindered progress. You needed no gift of second sight to see what had happened that night — the untidy and hurriedly discarded equipment cluttering up the trenches told its own tale. An urgent line of silent moving men pressing hurriedly along the trenches towards the beach — moving relentlessly like lemmings — stopping for nothing. The furtive discarding of their allotted burdens of blankets and equipment — the primal necessity to lighten their loads that night — the heavy loads that might impede a faster progress. To a soldier all these familiar signs told their own tale. It would have told the same story to a blind, deaf and dumb Indian tracker.

Stumbling over valises thrown away, piles of blankets left in all sorts of odd places, tripping over rifles left sprawled across the trench — to my fevered imagination it seemed to take me hours to negotiate even a few traverses. But the strangest, the most eerie and the most frightening element was the absence of all human life. The trenches — normally full of men and now deserted. It felt as if I were picking my panic-stricken way through the deserted catacombs of a dead and long-gone civilisation. Eerie! My God! I've never been so scared. The drifting heavy clouds over the moon's face alternate the night into an ever-changing pattern of light and dark — an unnerving night of sinister forebodings, where every little sound I made seemed like the clatter of Doom. In the course of a not very heroic life, I can honestly claim that that night I was the most frightened man the world's ever known or ever will know. Not even a rifle shot anywhere — to remind me that somewhere, even if only in the Turkish lines, there is still some human life left on this earth.

Seawards, too, the absence of all life. Not even the comforting view of the lighted hospital ship. I guess she's made her last trip here and is now on the way to Alex. Nothing to see, seawards — only the distant far-end of my world.

Round the next traverse of the trench I come across a soldier fast asleep on the firing step, head lolling on one side. Thank God at last for company, another human being besides myself on this nightmare planet. But it's the wrong sort of sleep. I shake him gently and he rolls flat on the trench floor, still and grotesque. There'll be no comforting chat with this chap. There's a bullet wound in his temple. He must have been killed very, very recently to have been left like that — probably in the mass press forward to the beach — probably killed through some silly sod not having adjusted the safety catch of his rifle properly. These things happen sometimes. They'd no time to even pretend to bury him — he'd been left where he was. The irony of Fate! Oddly enough, even his body was some slight comfort — proof, this momentary companionship that other beings besides myself had recently occupied this now dead and silent planet.

At last I leave the trenches and attain the open deep gully. Run soldier, run — you might still make it — you might still catch the last boat leaving the shore. I essay a few comforting words to myself — aloud — heartening myself with cheery optimism, but the voice isn't my voice at all. It sounds like someone miles away. Panting, swearing, falling flat on my face in the mud sometimes in my urgent haste I stumble along the gully and eventually reach the beach and join a few others stood on the edge of the shore and hollaring seawards. From out of the void of the night is a faint reply and now — God be praised — we espy a little pontoon-type raft being towed towards us. The raft already contains about a dozen men and we wade out and scramble aboard with them and the naval pinnace tows us slowly to the ship out at sea. The sea's a bit choppy and the raft alongside the ship bobs up and down dangerously. In the fitful darkness I am able to discern with difficulty a little hatchway in the side of the ship and one by one we wait until the heaving raft is level with the hatchway of the ship, then jump the intervening few feet separating us. It's a tricky job, and the naval bloke keeps bawling out hoarsely the need for extreme care. The chap who jumped before I did was real unlucky. His iron-studded army boots slipped on the wet raft as he jumped and he slipped into the water and as the raft bobbed up and down he got his skull crushed between the raft and the ship's side. My turn now — a panic-stricken long-jump record broken and I land stunned but safe on the iron footplates in the bowels of the ship.

CHAPTER 30

Exeunt

It was a good hour or so before the ship finally pulled out — an hour of waiting in case any more stragglers turned up. Give the British Navy its due — when it does a job it does it thoroughly. The rest of our Company had gathered together in small groups, subdued and quiet. The last 48 hours had been enough to keep any man quiet for life.

I found myself squatting alone at the stern of the ship as she pulled out, gazing landwards where an occasional gun flash and the fitful moonlight illumined the land we had just left. A figure emerged out of the dark and sat beside me. It was our Platoon Sergeant. We sat there quiet — not saying a word. I was imagining the Peninsula as it would be if we could see it properly — getting smaller and smaller as it would have been receding out of sight. Then the clouds would flit over the moon's face and we couldn't see damn-all except an occasional gun flash.

The Sergeant chuckled. 'Old Abdul'll get the surprise of his life in the morning. Not a bloody British soldier there.' He was quiet for a time, then in a voice so low it was almost a whisper, 'Except for the bloody thousands we've left there, Cobber. An' they won't be givin' Abdul much trouble, Ah'm thinking.'

I guess it was the reaction of the last few hours, but I found myself wracked with unmanly sobbing. I'm not sure why. I don't think it was so much the thought of all the chaps I'd known as kids and now left there — all the thousands of the other blokes who'd never breathe the air again. I don't know what it was — it could have been grief. But most, I think, was the dreadful feeling of the shame of it all — the British Army having to evacuate the Peninsula like this, and after all this gigantic wasted effort. Such things hurt a soldier more than you'd think — even an amateur soldier like myself.

'Wot's up, Cobber?' — and the older man slaps a friendly and comforting hand on my shoulder.

'Well, Sarge — all this — being kicked off the Pen. like this — us — the British Army.'

'Ah, cheer up, lad. He who fights and runs away lives to fight another day. Tha knows the old saying, Cobber?'

'Ay, I suppose you're right. But it does seem bloody awful, doesn't it?'

'Ah! Forget it, lad. Forget it. Cheer thiself up, land. Think of all them bints in Cairo and Alex. waiting for us.' And in a spasm of anger I rounded on him quite unjustly, too, and reverting to my native dialect as I always did when I got wound up, and using language quite unsuitable to a senior NCO.

'Thee and thi bloody bints, tha randy old bastard. All them lads of ours out there — and all tha can think about is them whoring bints in Cairo and Alex.'

He didn't get angry. Just kept quiet for a bit. In a low voice, more like a man meditating or something, he said:

'An' doesn't ta think, lad, that Ah feels just the same as thee. Doesn't ta reckon Ah knows wot it feels like to be leaving behind all them youngsters wot Ah've looked after ever sin they joined up — and won't be taking back with us no more. Doesn't ta think, lad, it's better to force thiself to think of bints, and booze and suchlike, instead of letting thi mind dwell, like, on all the things we've seen together, thee and me, lad. Or does ta prefer to let thiself go round the bend thinkin' of all the things we're a bloody sight better off for not thinkin' about? Doesn't ta think so, Cobber?'

Sometimes these army sergeants talk a lot of sense.

We sat there for about an hour, gazing back at the land we had just left — huddled in our greatcoats against the occasional splash of the sea spray. The Sergeant fishes out of his pocket some fags and we finish the packet between us 'and not saying nowt,' just looking back at the Peninsula — two insignificant specks of the human race alone with their own thoughts. Two insignificant strands of human cotton-waste spewed out from the mills of Lancashire who had been caught up in this drama of the Dardanelles as the loom of destiny weaves its own intricate and unpredictable future pattern.

The ship rolled a bit as it gathered speed. Below where we sat, only the rhythmic thump of the ship's engines obtruded the quiet of the night.

A reunion of No. 8 Platoon, 1/6th Lancs Fusiliers, at Rochdale Town Hall in 1938.

Seated in the front row is the mayor of Rochdale, Mary Duckworth, mother of Second Lieutenant Eric Duckworth, who commanded this platoon until he was killed in action on 7 August 1915.

Seated second from left in the front row is Charles Watkins.

APPENDIX I

Charles Watkins, a brief biography

Charles Watkins was born on November 3rd, 1895, in Nechells, Birmingham. He was the second of four children to Charles Henry Watkins, a clerk from Lichfield, Staffordshire, born around 1870, and Mary Ann Connor, born around 1863 in Aston, Birmingham. His siblings were Eva Mary, born on 27 July 1894, Frederick James, born on 7 March 1897, and Doris Kathleen, born in 1899.

The Watkins family had, by 1899, moved north to Rochdale in Lancashire. They were not wealthy but had some means. The 1901 census shows that they had a live-in domestic servant.

Charles Watkins attended Rochdale Secondary School from 1909. He left school at 15 to take up employment as a card room hand at the Turner Brothers cotton mill in Spotland, Rochdale.

When 17 years of age, Watkins volunteered for the Territorial Force, joining the 1/6th Battalion Lancashire Fusiliers on 7 May 1913. At enlistment, Watkins stood five feet, ten-and-a-half inches tall, with a chest measurement of 37 inches, above average in terms of physical stature compared to other soldiers at that time.

The 6th Battalion was headquartered in Rochdale. The town raised four companies, while nearby Todmorden and Middleton each provided two more. Like most locally-raised Territorial battalions, the men formed a close-knit unit. The battalion was commanded by the local lord, whom Watkins greatly admired.

> Lord Rochdale was the colourful figure of the town. He had been in The South African War commanding a local unit of yeomanry. Everybody respected him and liked him and he was more well known than a member of Parliament. [28]

In Watkins' opinion, he had the look of a man who liked his port wine.

> He had that sort of complexion. He had a moustache, twinkling eyes. All together a very lovable figure. [29]

Territorial battalions would parade on weekday evenings and Saturday afternoons to carry out drill and range practice. There were also weekends away on regimental exercises, and the much-anticipated annual camp with the division. For many 'Terriers,' who worked long hours in arduous employment, the twos-week-long camp under canvas was the highlight of their year. Watkins thought it akin to a summer holiday.

> It wasn't training, it was more a collection of getting the blokes together to get used to tent life. [30]

When war broke out in August 1914, Britain's regular army was well-equipped and trained, but tiny compared to Germany's. As it would be many months before Kitchener's New Army was ready to fight, Britain turned to her Terriers.

The Territorial Force had been raised for home defence only. However, many Terriers volunteered to serve overseas for the duration of the war, signing Army Form E.624, the Imperial Service Obligation. Among them was Private 8878 Charles Watkins.

On September 10, 1914, the 1/6th Battalion Lancashire Fusiliers embarked for Egypt with the East Lancashire Division (later designated the 42nd Division). They were the first Territorial division to be sent overseas.

---

28 Interview with Peter Liddle, February 1973.
29 ibid.
30 ibid.

A group of the "Terriers."

'Each day since the mobilisation orders were given the Rochdale Territorials have had to report themselves at headquarters and the equipment has been examined. Our photo shows a group of the men waiting outside the Drill hall for the nine o'clock parade on Thursday morning.'

'On Thursday afternoon the Territorials marched down company by company in order to have their bayonets sharpened... over 500 bayonets were prepared. Our photo shows the bayonets being polished after sharpening and one or two Territorials testing the keenness of the edge.'

The Rochdale Observer, 8 August 1914.

'Stepping out pluckily.'

'The Rochdale Territorials made an early move on Thursday morning, leaving their headquarters at five o'clock. They marched very well to their camp at Turton, about 15 miles away, and comparatively few men fell out. Our photo, which was taken a few miles beyond Bury, shows the column "going strong."'

'About every hour there was a halt for a few minutes and the men took the opportunity for a well-earned rest. At the second halt, just out of Bury, the Terriers ate some of the food they had brought in their haversacks, as shown in our photo.'

The Rochdale Observer, 22 August 1914.

For Watkins, war was a welcome escape.

> I left school at the age of 15, to go and work in a foul Lancashire cotton mill, 12 hours a day. As a Territorial soldier (volunteer) I was mobilised at the outbreak of War One on August 4th 1914, and went abroad to the Middle East Theatre of war at once. So the war itself was for me, and for most of us, a deliverance from slavery, and as such was a welcome relief.[31]

Watkins landed in Egypt on 26 September 1914. He settled into barracks with the battalion at Abbassia, on the outskirts of Cairo, and found a strange new world to explore.

Among the exotic sights were Australian troops, a popular type in Watkins' stories.

> The impact they made on all of us I think was a powerful physique and their colourful feathered hats and their complete indiscipline. A private would go out boozing with his captain and they would get drunk together arm in arm. We were brought up on the British army tradition of if it shocked us it somehow fascinated us. It spoke of a life that we would like to be but daren't be.[32]

For the next seven months, the battalion undertook training and route marches in the desert, or were assigned guard duties at various points around Cairo. Then, with little warning, on the first day of May 1915, came an order to entrain for Alexandria. From the railway siding that ran to the quay, the 6th Battalion marched up the gangway of the transport SS *Nile*. At 5.30 p.m. on May 2nd, they embarked for Gallipoli.

The Territorials were being sent as reinforcements to the Mediterranean Expeditionary Force. The regular army, who had landed on the Gallipoli Peninsula on 25 April 1915, were held up by the Turks' brave and stubborn defence.

---

31 Letter to Col. Mahakian, 7 September 1985.
32 Interview with Peter Liddle, February 1973.

On May 5th, the 6th Battalion landed at W Beach. It was the day before the start of the Second Battle of Krithia.

> You must appreciate that when we landed none of us had the faintest clue what we were going to do. Nobody told us anything. We were just there and we hadn't the faintest idea what to expect or what to do.
>
> So we just pushed our way forward and then we came under fire. Many officers were quite inexperienced at war as we were tried to get us to make some sort of line which we did and we came under very heavy rifle and machine gun fire and of course, shellfire was there all the time and we did what we could to retaliate but I must say that we did a lot of firing but we saw precious few Turks. They were well and truly hidden.

In 1973, Watkins was interviewed about his Gallipoli experience.[33] What were his thoughts as he waited to go over the top? Was he thinking of his mother, or his god?

> No, to be honest. All I thought about was that I hoped to God that I didn't get hit and if I do get hit I hope it is some part that is not vital...
>
> Because I am that sort of chap, I feared more than anything else being hit in the testicles.

Did he feel enmity towards the Turk?

> I didn't feel any resentment, I don't think any of the other chaps did. Any more than you felt resentment against the shells that might kill you or the rifle fire. It was a thing that you accept.

The 6th Battalion would endure until the closing of the fight. They were evacuated from Helles in the early hours of December 29th, 1915. They had lost 12 officers and 196 other ranks killed or died of wounds, and suffered, according to some estimates, between 700 and 800 wounded.

---

33 Interview with Peter Liddle.

After a short stay at Mudros, the battalion returned to Egypt for a period of rest and recovery. Later, they would fight in the Sinai desert, then on the Western Front. But Watkins remained in Egypt, where he entered the Royal Flying Corps (RFC) as Air Mechanic 2nd Class on September 2nd, 1916. His new service number was 403846.

At first, Watkins was placed in the administrative rather than the technical branch of the air service. A personnel record gives his qualifications as 'Associate of Arts knowledge of French and Arabic, bookkeeping, shorthand, typing,' and describes his work as 'organisational and administration, brigade office. Also general staff work, clerical work.'

On January 1st, 1917, Watkins was appointed Air Mechanic 1st Class, although downgraded to 2/AM six months later, on 23 July. Then, on October 1st, 1917, he was appointed Acting Corporal and regained his rank of 1/AM. On December 1st, 1917, he was promoted Corporal. When the Royal Air Force (RAF) absorbed the RFC on 1 April 1918, Watkins was transferred to the RAF as Corporal clerk.

Watkins' rise soon took a more literal form. As a Flight Cadet, he attended a balloon observer's course in Cairo, graduating on June 8th, 1918.

Kite balloons were used extensively by the air service for artillery spotting, observation and reconnaissance. The balloons were secured to the ground by a tethered cable, with a winch to control their ascent and descent. The sausage-like craft could reach altitudes of 5,000 feet and more, providing the observers with a wide-ranging view of the surrounding country. They worked from a basket suspended directly beneath the balloon, and communicated with the ground by telephone. Parachutes were carried in the basket, allowing the observers to jump clear of the balloon in the event of an attack or imminent danger.

Watkins was granted a temporary commission as 2nd Lieutenant (Kite Balloons) on 27 July 1918. In mid-September he was sent to the British Salonika Army on the Macedonian Front. He joined 22 Balloon Company, 16 Wing RAF, as an observer.

By this time, air superiority, and the initiative, had passed to the Allies. As part of a great attack along the front, the British assaulted Bulgarian positions in the hills above Doiran Lake on 18 September 1918. Watkins may have arrived just in time for the battle.

The British attack failed, but successes elsewhere by French, Serb and Greek forces caused the Bulgarian army to retreat. By the end of the month, Bulgaria was out of the war.

Watkins includes a vignette of life as a balloon observer in chapter 26. It was very different from Gallipoli.

> As a Tommy on Gallipoli, it was a worm's eye view. But now it is the lofty view from the air — the view of the bird of prey.

Watkins related, late in life, that he was pressed to write of his experiences in Macedonia.

> Someday I might get round to telling this story, always being assured by the military types that the memories of a Balloon Observer would be unique and valuable. I touch lightly on my balloon duties in Lost Endeavour, but the story itself is worth a separate book. What I really need is someone to twist my arm — or give me a good kick up the arse — to get me going on my balloon book, and who knows, I might creep out of my accursed lethargy and start boring once again my long-suffering readers of an almost completely forgotten World War One.
>
> But in my 90th year, I am, as they say in England, 'not this year's rhubarb' and I am getting lazier every year.[34]

Sadly, Watkins never wrote his balloon book.

> For me it was the most enjoyable time of the war — so much so that I could have wept when the war ended, for I loved the job so much.[35]

On November 15th, 1918, Watkins was ordered home. Five months later, he was transferred by the Air Ministry to the unemployed list.

Watkins lost his mother Mary Ann in 1918, perhaps before he had returned to England. His father, then aged 49 and working as a chartered accountant, remarried in 1919. His second wife was Bertha Williams, a 47-year-old art dealer.

---

34 Letter to Col. Mahakian, 7 September 1985.
35 Letter to Col. Mahakian.

By 1921, Watkins had moved to Kent and was running a laundry business. He was 25 years of age, and shared a house in Canterbury with his older sister, Eva, who was a teacher at the Littlebourne elementary school.

He met a local girl and married her in 1922. Mary Angelus Ellen Drew was born on 27 November 1904 in Medway, Kent. Her father had soldiered in India and South Africa.

The couple had four children, Doreen Mary (1923–2007), Cynthia Geraldine Mary (1928–2008), Charles M.F. (1936–1960) and Angela (1946–2006).

By 1939, Watkins had moved to Portsmouth. The census return shows that he had rekindled his love affair with the air force, although as a civil servant. Watkins was employed by the Air Ministry as an Air Service Clerk.

In 1943, Watkins' daughter Doreen (sometimes spelled Dorine) married William John Scull (born 1915), a mathematician in the Air Ministry, but he died just three years later, in 1946. Doreen remarried in 1949. Richard Stanton-Jones (1926–1991) was chief designer at the British aero-engineering firm, Saunders-Roe, where he developed ballistic rockets, hovercraft and amphibious landing craft. He later became managing director of the British Hovercraft Corporation and vice-chairman of Westland Helicopters. 'Dick' and Doreen had one son, Richard Stanton-Jones Jnr, who was born in 1950. The younger Richard was probably the only grandchild to Charles and Mary Watkins, as it seems that none of their other children got married.

By 1946, Watkins had moved north to Clapham, a village in the borough of Bedford. The move may have been prompted by his employment at RAF Cardington, an airbase five miles south of Bedford.

A holiday to Spain in 1967, and a chance conversation with an old man who had soldiered with the Spanish Legion, was the spur to Watkins writing down his 'sketchy reminiscences.' He tells the story in the preface to this book.

A draft of *Lost Endeavour* was completed in 1970, when Watkins sought the approval of Lord Rochdale's family.

> Courtesy seemed to demand that I made sure that nothing in my own book might offend the House of Rochdale, so I sent a draft

copy of the book to Lord Rochdale's son, who, on the death of his father became Viscount Rochdale. I offered to delete anything he might find not quite proper. He urged me not to delete a single word of it, for he had checked from his own deceased father's notes, much of what I had written. He also volunteered a foreword to the book itself.[36]

Watkins sent a copy of the book to the *Rochdale Observer*, whose editor replied with the following letter.

Dear Mr Watkins

I have read your book with intense interest. First, let me tell you what it is not.

It is not tightly and crisply written (indeed, in places it screams aloud for the attention of a ruthless sub-editor).

Now let me tell you what it is.

In my opinion, it is one of the most moving pieces of writing about Gallipoli that I have ever read. It is a book that needs to have been written in just your style. It would be ruined by being given the polish which a skilled sub-editor would introduce because then it would lack the conviction which is conveyed by the occasional stumbling sentence when a non-professional is struggling to put his sincerity on paper.

In other words, I raise my hat to you…[37]

In 1969, a society for veterans of the Gallipoli Campaign was established and Watkins was an early and active member (number 167). The Gallipoli Association aimed to 'keep alive the memory of the campaign among those who served in it.'

Fifteen extracts from *Lost Endeavour* were printed in the association's journal. The first extract appeared in 1970, in issue number 3.

---

36 Letter to Col. Mahakian.
37 Letter from N.F. Thornton, Editor-in-Chief, *Rochdale Observer*, dated 23 Dec [1970?].

The editor of *The Gallipolian* informed members that the book was available from the author, who had produced, at his own expense, a limited number of copies.

> There will be a small (less than cost) charge for this book. But it is definitely not a Book at Bedtime. Indeed, some of the episodes require a strong stomach, thanks to the author's insistence on absolute reality — as he saw it — (and even to the language at times!) But the author's compassionate treatment and understanding of the soldier more than compensates for the book's less salubrious episodes.[38]

Twelve years later, in 1982, the editor advised that a second print run was available.

> In response to demand Watty has had the complete book published and available on sale. It is the complete original, with some army language unexpurgated. It will cost about £10 plus post.[39]

Incidentally, the second edition of *Lost Endeavour* — 'produced for a limited and private circulation only' — was printed by Promotion House Ltd in Kent, the same printers responsible for *The Gallipolian*.

By printing extracts from *Lost Endeavour*, the Gallipoli Association helped bring Watkins' stories to a wider audience.

The Australian military historian John Laffin made extensive use of *The Gallipolian* when compiling veterans' accounts for his 1980 book *Damn the Dardanelles!* Among these personal accounts are four attributed to 'my friend' Charles Watkins, 'the large-eyed, big-eared little Lancashire man.'

Laffin had a high regard for Watkins' writing.

> His description of life at Helles is one of the most vivid in Gallipoli literature.[40]

---

38 *The Gallipolian*, issue 4, 1970.
39 *The Gallipolian*, issue 40, 1982.
40 John Laffin, *Damn the Dardanelles!*, p. 218.

Watkins was similarly enamoured of Laffin's book.

> "DAMN THE DARDANELLES" is, in my opinion, the best historic account of the whole campaign.[41]

Another 'customer' for *Lost Endeavour* was Colonel Carl K. Mahakian of Burbank, California. Mahakian was an officer in the US Marine Corps Reserve. Watkins responded to his enquiry in a letter dated 7 September 1985.

> Thank you for your letter of the 25th August. Let me say at once I am flattered and highly honoured that a Colonel of such a prestigious corps as the U.S.M.C.R. should take an interest in the valiant, but shamefully conducted campaign as was the Gallipoli campaign, with its tragic, insane waste of human heroism. But enough of my indignation.
>
> Let me also insist I am no author, nor have I any pretensions of being a writer…
>
> I first wrote my book on Gallipoli in the late 1960s (that was the book from which the 1970 quotation appeared in our own little Service Magazine 'The Gallipolian'). Some three years later I did a re-write of it in response to pressure, as the few copies of the original book were soon exhausted, so I polished the first book up a little in the new re-print. But the story itself beginning at Chapter I of the new re-print (the copy I am sending to you) is exactly the same as the first book. I merely added a Preface.
>
> I *think* you will like my little book. But whether you do, or do not, I am sure you will be impressed by its honesty, for it would be foolish of me to expect you to be impressed by any vainglorious words of mine. You, as a soldier will appreciate truth, even if it is at times rather unpalatable.

Colonel Carl Karnig Mahakian (1926–2015) served in World War II, the Korean War and the Vietnam War. He also worked in radio, television and film, winning two Emmy Awards. The US Marine Corps'

---

41 Letter to Col. Mahakian, 1 October 1985.

interest in the Gallipoli campaign is well known and may be the reason for Mahakian's interest in Watkins' book. However, his extensive work in Hollywood does raise an intriguing thought... was Mahakian planning to pitch a film about the campaign?

As well as providing the Gallipoli Association with excerpts from *Lost Endeavour*, Watkins also wrote a few original pieces for their journal. He attended annual dinners held by the association on at least two occasions, in 1976 and 1979, and reported on the former as 'idle thoughts by an idle fellow' (issue 22, 1976). He wrote an appreciation of the association's membership list (issue 8, 1972) and remembered 'our ghostly guests' (issue 25, 1977).

Watkins also provided a short piece on the oak tree planted in Redoubt Cemetery by the parents of 2nd Lieutenant Eric Duckworth (issue 6, 1971). Watkins is quoted extensively in a fine modern biography of Duckworth that tells the story of this unique memorial and the 1/6th Battalion Lancashire Fusiliers at Gallipoli. Readers are encouraged to seek out *The Gallipoli Oak* by Martin Purdy and Ian Dawson.

Watkins survived his wife, who passed away in Bedford on 23 October 1985. He moved to the Isle of Wight in 1988 to live with his daughter Doreen and her husband, Dr Richard Stanton-Jones, at 'Doubloon,' Oakhill Road, Seaview. He died there, aged 93, on January 5th, 1989.

Brigadier General Herbert Cokayne Frith, CB.

APPENDIX II

Notes on individuals mentioned in the text

'The Splendid Brigadier' — Brigadier General Herbert Cokayne Frith, CB

CHAPTER 1

There's no denying it — he was a splendid figure of a man. Everything about him gleamed and shone — from the peak of his cap, heavily splashed with the 'scrambled egg' of rank, right down to the toes of his highly polished boots and leggings. His square-toed boots fitted snugly in the gleaming stirrup irons. Even the accoutrements of his mount, a magnificent chestnut mare — all the jingling harness, the snaffle chains, the bit, the reins and the saddle — everything shone and sparkled in the morning sunlight. The restless mare tossed her disdainful head as she faced us, giving the Brigadier a chance to display his superb horsemanship ... Oh, yes! We were properly impressed — as indeed we were meant to be.

300 years of breeding and family sat in that saddle — a long heritage of good schools, tradition and impeccable breeding looking down on us all, gathered there in sullen silence in the gully ...

Herbert Cokayne Frith was born on 12 July 1861 in Gainsborough, Lincolnshire. He was commissioned into the Argyle and Bute Artillery (Militia) in 1881, before joining the 2nd Bn Somersetshire Light Infantry as a lieutenant on 28 January 1882. He was promoted to captain on 23 September 1887.

Frith was employed with the Egyptian Army from 23 September 1885 to 4 October 1895, serving in the Frontier Field Force in Soudan 1885–86 and was at the Action of Giniss, being awarded the Bronze (Khedive) Star. In 1889, he took part in the Action of Toski (Soudan) and was awarded the clasp and also 4th Class Medjidie.

Frith was posted to the Indian Army as station staff officer class 1 (Punjab district) on 28 July 1897, until 28 July 1902, and was promoted to major on 27 January 1900. Listed as a qualified interpreter in Arabic in October 1902, Frith was also fluent in Turkish, French, Persian, Urdu and Pushtu.

On 27 November 1909, he was promoted to lieutenant colonel and took command of the 1st Bn Somerset Light Infantry. He was promoted to colonel on 2 June 1913.

Frith relinquished command of his battalion on 27 November 1913 to take command of the Lancashire Fusiliers (125th) Brigade. He was listed as temporary brigadier general on 5 August 1914.

Frith commanded the brigade in Egypt and arrived with it at Helles on 5 May 1915, commanding it throughout the campaign. He was made temporary major general on 30 December 1915, until 20 January 1916. He was awarded the CB in 1916 and was twice Mentioned in Despatches for his service on Gallipoli (LG 5 November 1915 and 13 July 1916).

Frith commanded the 125th Brigade in Egypt in 1916–17 and moved with it to France in March 1917. He returned to England on 23 June 1917 to take up command of the Home Service Brigade, an event commemorated in the divisional history:

> General Frith was the last of the General Officers who had served with the Division from the outbreak of war. For three years he had commanded the Lancashire Fusilier Brigade, which had become much attached to him, for he was quick to recognize and give credit for good work, and he possessed a remarkable memory for faces, invariably knowing each officer by name from first meeting.

Frith was awarded the Order of St. Stanislas, 2nd Class with Swords, and retired on 1 November 1918. He died on 5 March 1942.

INDIVIDUALS 231

Lieut. Colonel Lord Rochdale (George Kemp)

CHAPTERS 3, 5, 6 AND 15

I am a staunch and rude democrat; they don't come ruder than me, but for this particular member of our aristocracy and for the many others like him in the two World Wars, I am the biggest lickspittle of them all.

I found myself that first day, and by pure chance, right alongside him as we scrambled up the low cliffs in twos, threes and fours in this broken country. Woefully short of breath and with the sweat streaming down his face, revolver in one hand and walking stick in the other, urging his men for'ard — this Lord Rochdale. By rights, and at his age and physical condition, he should very properly have been occupying a safe seat at the War Office or at some HQ. Base, and where he could have been and without the slightest loss of honour, instead of being on this young man's Commando-type enterprise. But not for him the soft seat of war. Like many others of his kind, he just wasn't that sort. And he stuck with us during the campaign, sharing equally our dangers and hardships until age and its infirmities finally caught up with him.

George Kemp, born on 9 June 1866 in Rochdale, received his education at Mill Hill and Shrewsbury schools before pursuing studies at Balliol College, Oxford, and eventually Trinity College, Cambridge. On leaving university he entered the family business of Kelsall and Kemp (flannel manufacturers) and pursued an interest in politics. He was commissioned as 2nd lieutenant in to the Duke of Lancaster's Own Yeomanry on 3 October 1888 and promoted to captain on 18 July 1891. As the Unionist Party candidate, he was elected MP for Heywood in the 1895 election and retained his seat in the autumn 1900 election (while serving in South Africa 1900–1902).

Kemp served in the South African War with the Duke of Lancaster's Own Yeomanry, and was promoted to major on 20 September 1901. He was placed in command of the 32nd Bn Imperial Yeomanry on 13 January 1902 and promoted to lieutenant colonel (temporary rank in the army) on 22 January 1902. He was twice Mentioned in Despatches (LG 8 February and September 1901) and received the Queen's medal with two clasps.

Lieut. Colonel Lord Rochdale.

Kemp resigned his commission on 22 January 1904 to devote himself to the family business. He was knighted in the King's Birthday Honours list in 1909. In 1910, Sir George Kemp was again elected MP, this time as Liberal candidate for North West Manchester, but resigned his seat in 1912 due to his objections to certain Liberal Party policies. He was made Lord Rochdale in the King's Birthday Honours list in 1913 for his services to politics. With the launch of the Territorial Army, Rochdale became chairman of the East Lancashire Territorial Association and, on 13 December 1909, took command of the 1/6th Lancashire Fusiliers. At the outbreak of war, Lord Rochdale, by now 48, volunteered for overseas service and went to Egypt as the CO of the 1/6th Lancashire Fusiliers.

Rochdale landed with his battalion at Helles on 5 May 1915, the day prior to the Second Battle of Krithia. The following day he led his battalion in an attack on Gully Spur, displaying great personal courage as he encouraged his inexperienced troops. As one on the division's most experienced commanding officers, he was chosen to be temporary GOC of the 126th Brigade, from 21 May to 3 June, and temporary GOC of the 127th Brigade from 4 to 21 June 1915 (Rochdale taking over from Lt. Colonel Heys who had himself taken temporary command of the brigade when Brigadier General Lee was wounded).

Rochdale was a vociferous critic of the way the campaign was being conducted, and, using his House of Lords privilege (and against the advice of his superiors), he left Helles for England in early July. On 26 July 1915, he met with Prime Minister, Herbert Henry Asquith, First Lord of the Admiralty, Arthur Balfour, and Secretary of State for the Colonies (and leader of the opposition), Andrew Bonar Law, and submitted a detailed and highly critical report, fully aware of the damage he was doing to his career. He returned to Gallipoli in early August and was again placed in temporary command of the 127th Brigade, from 19 to 26 August. Rochdale was evacuated from the peninsula on 29 September 1915, suffering from para-typhoid and phlebitis. He never again commanded on active service and after recovering went onto the Reserve of Officers list.

Despite Rochdale's significant contributions in the campaign, and the many instances of personal bravery, these remained unrecognised, no doubt as a consequence of his actions in July, and it shows a shocking vindictiveness within the award process that he did not even receive a Mention in Despatches. He died in Lingholm near Keswick, Cumberland, on 24 March 1945.

Lord Rochdale (left) with a staff officer on the island of Imbros, 1915.

'Jack Gibney'

CHAPTER 3

... if Jimmy [the 'cocky little white-haired fox terrier'] *belonged in part to all of us, he belonged by common consent in particular to one Jack Gibney. Jack was the one the little dog always sought out among the crowd of us lounging in the evening on the barrack square, deftly and good-humouredly dodging our good-natured kicks as he roamed amongst us seeking his rightful master. A quiet, rather inarticulate lad was Jack, and not over-given to conversation, but he was the adored god to this little tyke ... so it was particularly hard on Jack that he should have to be told that his pal had to walk the plank ...*

But Jack needn't have worried all that much. I stumbled across him two or three days later on the Peninsula. He was lying stretched out, his eyes wide open to the noon-day sun. His body had the rigidity of the final sleep, but his face bore the relaxed, happy expression of a man who has at last whistled up his lost dog.

Probably Pte 8620 John McGibney, killed in action 15 May 1915, age 20. John was born in Castleton (half way between Rochdale and Middleton) in 1895. He was single and prior to the outbreak of war lived with his widowed mother, Mary, and was employed as a paper tube maker. John enlisted at Rochdale and judging by his service number had been a pre-war Territorial with at least two years' service (Watkins joined the battalion on 7 May 1913, and his service number was 8878).

John McGibney has no known grave and is commemorated on the Helles Memorial, panel 68.

The old sweat, 'Cpl I——'

CHAPTER 9

He was — par excellence — all that we imagined a soldier should be — tough, brawny and a hard drinker. Hand-carved by Kipling, this bloke.

In addition, he was the most tattooed man I'd ever seen. Specimens of the tattooists' art in India adorned most of his body... but the 'piece-de-resistance' was a hunting scene, the horses on his left breast galloping over his shoulder and following a pack of hounds that trailed down his back, chasing the elusive fox...

Watkins provides a great deal of information on this character who he describes as an old sweat, 'with the ribbons of the South African War and other campaigns,' an 'ex-regular of some Cavalry Regiment, veteran of many a skirmish of the Khyber Pass District,' and who had retired with 21 years regular service in the Army. Watkins also tells us Cpl I—— joined the 1/6th Lancs Fusiliers in 1913 (the same year as Watkins) and that he was single and had never married. Despite these clues, which point to the man being in his 40s with an army number in the mid-8000s to early 9000s, the editors have been unable to find a single possible candidate amongst the fallen of the battalion (for the entire campaign). For now, Cpl I—— must remain an enigma.

'Sergt R——', killed on the same night as 'Cpl I——'

CHAPTER 9

Three sergeants died during or shortly after the Second Battle of Krithia: Sgt 7079 James Burgess and Sgt 7554 Henry Webster were both killed in action on 8 May 1915, while Sgt 7273 William Thomas Nelson died of wounds on 15 May. Any one of these could be Sergt R——.

'Frank Whitfield'

CHAPTER 11

A close boyhood friend of mine, Frank Whitfield, was next to me as we went over the top and a few yards out was shot in the temple ... On leave in my home town just after the war, I was astonished to bump into him, alive and well, and with three pips on his shoulder and the RFC pilot's wings. Miraculously he'd survived his early wound and with no ill effects.

Probably Sgt 8613 Francis Whitfield, 1/6th Lancs Fusiliers. Judging by his service number Francis had been a pre-war Territorial with at least two years' service (Watkins joined the battalion on 7 May 1913, and his service number was 8878).

Commissioned Sub Lieutenant in the Royal Naval Volunteer Reserve on 30 April 1918. Transferred to the Royal Air Force on 8 September 1918 but returned to the Regimental Depot of 63rd (Royal Naval) Division at Alnwick in May 1919 where he was demobilised on 28 May 1919.

'Cpl Grimes'

CHAPTERS 12 AND 13

One of our chaps, Cpl Grimes, a married man with two kids, always kept a woman's worn cotton stocking folded on top of his belongings. At times we'd observe him finger it longingly but we tried hard not to see this damned stocking. The rougher soldiers are the more delicate they are not to pry into a man's sentimental privacy, especially domestic privacy. Even scallywags like us, to whom nothing and nobody was sacred, would respect a man's domestic privacy and the reputation of a man's wife. If we thought about that damned stocking at all, and we tried hard not to — it might have been a tenuous link with his wife, a memento of other happier days. Nothing unusual in that — to us. We Lancastrians allow sentiment to have an unusual place in life and are openly unashamed of it.

During the usual sultry morning natter one day about this and that, a gust of wind dislodged this damned stocking, depositing it near the feet of one of the young lads nearby. With an air of awkward unconcern, he picked it up and gave it to its owner. Cpl Grimes thanked him gravely then burst out laughing.

'Did Ah ever tell thee, lads, 'ow Ah got this bloody stocking? ...'

This is another character that the editors have searched long and hard for his true identity, but have been unable to find a clear-cut candidate amongst the fallen of the battalion (for the entire campaign).

Watkins clearly sets this episode during the short time the 1/6th Lancs Fusiliers occupied the firing-line at Fusilier Bluff (nine days between 3 September to 17 September). During that time only three men of the 1/6th Lancs Fusiliers died.

Pte 9487 Harold Lees was killed in action on 3 September but is ruled out by being single. Pte 9983 Samuel Duckworth was killed in action on 7 September and was married with two children but is also ruled out as he arrived on the peninsula with a draft on 22 July, that arrived directly from England, and as such he had no opportunity to experience the 'delights' of Egypt. Cpl (L/Sgt) 9273 Joseph Marsden died on 10 September and although married with two children can also be ruled out as he died of dysentery in hospital in Alexandria. As such it is probably safe to assume the story is either an amalgam of memories or completely apocryphal.

'Fred Buckley'

CHAPTERS 15

We had in our gang one Fred Buckley whose sly wit was a constant joy to us ... He seemed for ever to be on the look-out for someone he could give a bit of practical help to ...

One day, and while temporarily deserting his cook-house duties, he was with us in the trench when a hand-grenade attack suddenly materialised from Johnny Turk. Instead of nipping back to his proper duties in the cook-house, Buckley joined gleefully in the fray until an exploding grenade tore out his left shoulder. I never knew what happened to him after he was carted away, groaning, on an improvised stretcher, or whether he lived or died. For me he still lives.

Probably Pte 10190 (240949) Fred William Buckley, who was 21 or 22 when he arrived at Gallipoli with the draft of 21 June 1915. A native of Middleton, in civilian life Fred worked as a cotton mule piecer and was single man living with his parents, Clara and John. He was still serving with the 1/6th Lancs Fusiliers when he died of illness in France on 20 March 1918. He is buried at Longueness (St Omer) Souvenir Cemetery (VI. F. 72). His headstone carries the poignant inscription: 'We link our hearts to thee — Until we meet again — Dearly loved.'

The 'Machine Gun Sergeant' and the 'Machine Gun Corporal'

CHAPTER 16

... two or three days before the evacuation... the Turks were making a last desperate attempt to break through our front line. They were unlucky, but it was fierce while it lasted. The Machine Gun Sergeant was all het up and agitated. I was struggling to free a jammed belt in the machine gun.

'Can tha manage on thi own, Cobber?'

'Ay, I can manage, Sarge. Why, what's up?'

'Well, the lads is a bit pushed, like. Hundreds of the bastards trying to break through our line. Me and the Corporal thought as 'ow we'd go and lend 'em a 'and, like. Sure tha can manage?'

'Ay, I'll be all right. Good luck, Sarge.'

'Ay... well...' He paused, regarding me doubtfully. Eventually I hit on the right cuss words and got the belt free. Then the fusillade of shots grew louder and he and the Corporal picked up their rifles and dodged up the little sap leading from our gun emplacement to the front line to 'lend a bit of a 'and, like'. The Sergeant was a short chap, about 5' 2", and the Corporal a long lanky 6' 3". But the difference in height didn't seem all that noticeable when they were laid head to foot on the trench floor a few minutes later.

The sergeant is probably Sgt 7646 John L. Taylor Mellor. Although listed as being killed in action on 21 December 1915, Mellor could quite easily have been killed two days earlier (reports of casualties in battle were often delayed or inaccurately recorded). John was born in Bury and in civilian life had been employed as a piecer in a cotton spinning mill. Judging by his service number he had been a pre-war Territorial with at least two years' service (Watkins joined the battalion on 7 May 1913, and his service number was 8878) and had enlisted in Rochdale. He was 26 when he died and left a widow, Annie. John has no known grave and is commemorated on the Helles Memorial, panel 60.

The corporal could be any one of the five corporals listed in the Honour Roll (Appendix VIII) as being killed in action or died of wounds between 19 and 22 December 1915: Cpl 8278 Chadwick Davies, L/Cpl 9744 Samuel Ingham, L/Cpl 9581 Frank Turner, L/Cpl 8668 George Wild and L/Cpl 8315 William Scholes.

The men were probably casualties of a British counter-attack following a diversionary action on the 'Gridiron' on 19 December 1915. See Appendix II for a description of the diversionary actions at Helles on this date.

'Johnny Stirk'

CHAPTER 18

Johnny Stirk never got a chance to bash his cap in or even to blacken his buttons. No! It's not his real name, in fact I'd never seen him before, but it was 'Johnny' something or other — I forget his last name. Just arrived in our sector of the trench, bursting with eagerness and curiosity, wouldn't even wait to divest himself of his full pack, but must needs stick his head over the parapet. 'I must take a look at these Turks.' Too late to hear the warning cry of one of our chaps, 'Get down, tha silly young bugger, get down.' Before you could count three, his forehead was neatly drilled.

Probably another of Watkins' apocryphal characters. Of all the men who arrived with the various drafts none of those who were killed in action (see Appendix VIII, Honour Roll) match the comparative youth of 'Johnny Stirk' nor the circumstances of his death.

'Tom Sutcliffe'

CHAPTER 18

Tom Sutcliffe had been a 'piecer-up' before the war, at the Standard Mill ... In his usual lazy leisurely way of speaking, he painted the picture, at times good-humouredly, but mostly in a bitter undertone of these 'piecers-up', the young lads who toiled endlessly 10 hours a day in our cotton factories ...

'No, mates — give me soldiering every time. Even this bloody balls-up on the Peninsula, to goin' back to that life again.'

This cannot be the character's correct name. The only Tom Sutcliffe (Pte 9878 Tom Sutcliffe) to serve in the 1/6th Lancs Fusiliers during the period of the Gallipoli Campaign was a former miner who did not serve at Gallipoli.

'Cliff Stansfield'

CHAPTER 18 AND 23

'Course 'e ain't married,' said Cliff Stansfield decisively, and joining in the argument. 'Why! 'e don't even look *married,' asserted Cliff, reasonably enough. 'He looks 'appy.'*

Probably Pte 8824 (203410) Clifford Stansfield. He arrived on the peninsula on 5 May 1915 and later served with the 1/5th Lancs Fusiliers. Clifford survived the war and was disembodied on 7 March 1919.

The Cobbers: 'Two NCOs of the Australian Artillery'

CHAPTER 22

I was on a special and lousy, thankless duty one day in a little 'sap' (or small trench) abutting forward at right angles from our front line. At the end of this little sap — about 20 yards in front of our front line — was fixed a sniper's hidey-hole — an inch-thick steel plate with an aperture in it just large enough to poke a rifle through. You swivelled aside a little piece of the same inch-thick steel, and designed just large enough to overlap the aperture when not in use, poked your rifle through, took a quick sight at anything moving in the Turkish lines ahead, and try a quick shot, making sure you're back to safety and with the aperture closed again, and all this before you can count ten — otherwise you'd get one back through the aperture. For like I told you, these Turkish sharp-shooters are terrific...

Two NCOs of the Australian Artillery — huge chaps — about 6'4" and broad with it. They were seeking an Artillery observation post as far forward as possible for the next offensive of ours. That's how they came to be in front of our front line and chatting with me. One was a Staff Sergeant and did all the talking and the other, a Corporal, just listened and didn't say a lot. The Staff Sergeant grew intrigued with my job and wanted to have a go himself but I told him the score and that if you didn't know the trick, you'd be a dead duck before you could look round. This seemed to peeve him — 'didn't want any Pommie to tell him all about shooting'. 'Back in Austry-lia he could shoot the toe-nail off a 'possum's left foot'...

So I handed over, deeming it best to say no more, and stood alongside him pissing myself with anxiety while he took a long and careful sight. Then there was a 'plop' and when the Corporal and I picked him up he was as dead as ever a man can be. Then the Corporal cussed and swore — said 'I'll get the bastard who did that to my pal' and in spite of my protests, poked the rifle through the hole himself. I hollared for our Sergeant, who came quick. But not quick enough and the Corporal, badly wounded, was later carted off by the stretcher-bearers. The body of the Staff Sergeant was carried to our front trench and we sent word back to his Battery as to what had happened.

On 15 June 1915, when this episode probably occurred, the 1/6th Lancs Fusiliers were holding the Right Sub-section of the firing-line which stretched from Achi Baba Nullah on the right to the communication trench that ran along the western side of the Vineyard. The British firing-line ran along the southern end of the Vineyard and, before the advance of 4 June and the capture of the Turkish firing-line, several similar communication trenches ran back towards the Turkish support lines. These communication trenches were subsequently blocked with barricades and Watkins may well have been manning a loophole in one. Although his battalion had been relieved from the firing-line by the 1/8th Lancs Fusiliers around 3.30 p.m. on the previous day, Watkins may have remained behind temporarily with other 'snipers' to pass on hard-earned knowledge and operating procedures relating to the posts.

On this date, several Australian artillery batteries diverted from Anzac were emplaced at Helles. These were the 1st, 2nd and 3rd batteries of the 1st Field Artillery Brigade (FAB) and 6th Battery, 2nd FAB.

The war diary of Headquarters, 1st Australian Field Artillery Brigade, records the following casualties for 15 June 1915:

> Gr Pearson BHQ killed. Br Felstead 2nd Bty wounded.

The war diary also notes in its entry for 15 June that a party of sharpshooters was to 'take up positions in forward trenches for sniping purposes.'

Watkins gives the name of the staff sergeant as Jack Ballantyne, but the editors believe the unfortunate Anzac to be Gunner 3513 Stanley Pearson of Headquarters, 1st Field Artillery Brigade, AIF. Gunner Pearson was 5 feet 8 inches, not 6 feet 4 inches. He was born in London and deserted from the Royal Navy in 1913 when his ship was berthed in Sydney Harbour. He enlisted with the AIF in December 1914.

The wounded man, 'the Corporal,' was probably Bombardier 195 William Herbert 'Bill' Felstead, 2nd Battery, 1st Field Artillery Brigade, AIF. He was 20 years of age, close to six feet in height, with pre-war militia service with the Australian garrison artillery. Felstead later served on the Western Front and was granted the temporary rank of captain in June 1918 while commanding a trench mortar battery. He returned to Australia in May 1919 and was living in Redan Street, Mosman, in 1967. He died in 1971.

The clue to the identity of these men came from the Gallipoli diary of Alec Riley, who recorded the death of an Australian in the firing-line on the afternoon of 15 June 1915. The location and circumstances were a close match to Watkins' account. A review of all Australian artillerymen killed and commemorated at Helles supported the likelihood of Pearson being the gunner who was killed.

'One of our officers…'

CHAPTER 23

The trench was empty, empty except for one figure a few yards away — one of our officers … The appearance of this particular officer shocked me. Unshaven, haggard and bloodshot-eyed, it was difficult to recognise in him the one who was under all conditions the soldier 'par excellence.' His legendary smartness had never seemed to desert him ever since we had landed, and in the heat of previous battles he was forever of a parade-ground smartness that was a source of wonder and inspiration to us. When we would be mucky, battle-stained, dirty and unkempt, his leggings were always of the same beautiful polish, his uniform impeccable, and his face always scrubbed and clean. Like an invincible and inviolate god, he always appeared to us — by contrast. Nor was he ever flurried or flustered like us — always calm, efficient and superior to all conditions, even appearing always slightly amused by everything — like as if this sort of soldiering was some sort of game.

All the greater therefore was the shock when I met him, like a man who had been on a three days boozing jag and hadn't yet got over it. No battle hazards in the world could have reduced this officer to such a state — he just wasn't that sort of chap. Then I remembered whispered rumours that had been going around about him — rumours whispered so discreetly and sorrowfully by us, for no man enjoyed such respect and admiration as did this chap. So I began to wonder what sort of Dear John letter he might have received to have reduced him to such evident wild despair.

Only two 1/6th Lancs Fusiliers officers died around the time frame described by Watkins. One of them, Lieutenant Joshua Harold Smith who was killed in action on 12 August 1915 (buried in the Redoubt

Cemetery) was 24 years old and single so can be ruled out. The other was Lieutenant Frederick William Harvey, 10th South Lancs Regt (attached) who died of his wounds on the 9th or 10th of August 1915, and is buried in Lancashire Landing Cemetery. Frederick was 35 years old and married — these two facts and the date of his death concurring exactly with Watkins' tale. However, Fredrick came from a relatively humble background (the 1911 census has him living in a three-bedroomed house with his wife, their four young children and his wife's sister and gives his occupation as a travelling salesman for a flour mill). As such it seems unlikely that his wife would have had the means or the opportunity to travel to Cairo, leaving behind four young children.

Taking everything into account, it seems likely that Watkins' tale may have a grain of truth in it. That two days after the start of the battle, Watkins had watched an older and much-respected officer walk to his certain death. The rest of the story, of the wife's arrival in Cairo and subsequent infidelity, should be taken with a pinch of salt. Trenches and barrack rooms are the breeding ground of rumours and the editors know of several similarly spurious variations on this theme. In fact, very few officers of the East Lancs Division were joined by their wives in Egypt in 1914 or 1915.

Captain William Henry Griffiths, Captain Quartermaster, 1/6th Lancs Fusiliers

CHAPTER 24

A great bloke, our Battalion Captain Quartermaster. And this time there's no indiscretion in naming him — Captain William Henry Griffiths — South African War veteran, cool, caustic and clever.

Captain William Henry Griffiths, DCM, was far from your standard Territorial battalion quartermaster. These were typically former senior NCOs who had re-enlisted with the honorary rank of lieutenant after long service with the regulars. This was not the case with Griffiths who had seen much active service, and as a squadron commander, had led troops on campaign in the Second Boer War.

Born in Hoxton, Middlesex, on 9 May 1860, he was the son of William and Mary Griffiths. His father was a bookseller's clerk (and a private in the 6th Middlesex Volunteers).

Griffiths began his military career as a private in the 7th Dragoon Guards early in 1879 and served with the regiment in the Egyptian Expedition of 1882, earning the Bronze (Khedive) Star. By 1883, Griffiths and his regiment were at Shorncliffe Army Camp, Kent. In November 1883, he married his first wife, Annie Kate Belcher, who bore him a daughter, Mabel Anne, in 1886, and a son, William Harry, in 1888.

Griffiths continued to serve with his regiment, rising to the rank of squadron quartermaster sergeant. In 1897 he was posted to A Squadron (Oldham & Rochdale) Duke of Lancaster's Own Yeomanry, then under the command of Captain George Kemp MP (later to become Lord Rochdale), as a permanent staff instructor.

Early in the Second Boer War, between 10 and 17 December 1900, the British Army suffered a string of defeats later known as Black Week. These devastating setbacks led to a Royal Warrant being issued on 23 December 1900 allowing volunteer forces to serve in South Africa. Yeomanry units were asked to provide companies of 120 men, to operate as Mounted Infantry. Officers and men from A Squadron volunteered and were formed into the 23rd (Lancashire) Company, 8th Battalion Imperial Yeomanry, under the command of Captain Kemp with Griffiths being posted to the company as squadron sergeant major on 5 January 1900. The 8th Battalion landed in South Africa on 5 March 1900,[42] and in early May, was attached to Sir Charles Warren's Column operating in Griqualand. In the early hours of 30 May 1900, a Boer force surrounded the column as it bivouacked at Faber's Put. As the Boers prepared to attack, a Yeomanry sentry raised the alarm and opened fire. After a furious firefight, the men of the 8th Battalion went on the offensive and advanced in rushes over open ground, driving off the Boers with the points of their bayonets.

---

42 Griffiths' own record of his service in South Africa gives his date of embarkation from Liverpool as 11 February 1900, so it's possible that the 23rd Company arrived earlier than 5 March.

Thus began Griffiths' impressive service record in South Africa. On 12 December 1900, he was commissioned lieutenant (with the temporary rank of lieutenant in the army). On 23 March 1901, Griffiths was placed in command of the 23rd Company (following Captain Kemp's appointment as second in command to the 8th Battalion) but in the same month he received news from home of his wife Annie's death. He was promoted to captain (with the temporary rank of captain in the army) on 15 May 1901, completing his extraordinary progression from sergeant major to captain in five months. Griffiths was mentioned in Lord Robert's despatches on 1 September 1901, and on 27 September 1901 was awarded the Distinguished Conduct Medal for actions carried out while serving as the 23rd Company's squadron sergeant major.

The last of the Boers finally surrendered in May 1902 and the signing of the Treaty of Vereeniging ended the war on 31 May 1902. Appointed second in command of the 8th Battalion early in July 1902, Griffiths was recommended for promotion to major. The 8th Battalion embarked at Cape Town for England on 11 August and was disbanded at Aldershot on 7 September 1902.

After 23 years of meritorious and dedicated service to the Crown, this signalled an abrupt end not only to Griffiths' employment, but his way of life. He faced a very uncertain future. On 11 September 1902, Griffiths, having had notice of his imminent discharge, wrote letters to the offices of the Commander in Chief, Lord Roberts, and the Adjutant General, requesting 'suitable employment in accordance with the rank I already hold,' and again on 17 September, this time adding 'he was perfectly willing to accept any appointment, either abroad, at home, [or on] active service, the latter of w[hich I] would greatly prefer.' Despite his urgent pleas, Griffiths, with many other redundant officers of the Boer War, was given no option but to relinquish his commission on 1 October 1902. However, unlike most of those officers who had a profession to return to, or private means, Griffiths as a 'ranker' and in middle age, had nothing to soften his return to the civilian life he had left 23 years earlier.

With his letters becoming increasingly desperate, Griffiths continued to petition the War Office for employment. On 19 January 1903, his persistence was met with the award of 'a special rate of retired pay, viz five shillings a day,' and the ungenerous condition that this was 'on the distinct understanding that it is in lieu of all pension in respect of your service, and that your wife and children will not be eligible for either pension or compassionate allowances in the event of their surviving you.'

On 18 March 1903, Griffiths applied to be placed on the Reserve of Officers, which was approved and gazetted on 19 May 1903. This provided additional income and some financial security. On 16 April 1904, Griffiths married a second wife, Annie Louisa Oxley, who was 15 years his junior. He resigned his commission on 20 January 1905, but correspondence dated 22 December 1906 indicates Griffiths' intention to secure work at the War Office. The last such letter in Griffiths' file was addressed to 28 Great Russell Street, Bloomsbury, the same house he had moved into prior to his discharge. By the time of the 1911 census, Griffiths had moved to 'Stembridge,' a substantial house in Golders Green, London. His status was recorded as a 'retired army officer' with a wife, Annie, and two sons, George William, aged 4, and Bernard Harold, aged 1. The family also had a servant and a boarder living with them. Later that year, in a move that must have been at the invitation of Lord Rochdale, Griffiths returned to Rochdale and took up the post of battalion quartermaster for the 1/6th Lancashire Fusiliers.

At the outbreak of the First World War, Griffiths, by now 54, volunteered to serve overseas and went to Egypt with the 1/6th Lancashire Fusiliers. He landed with the battalion at V Beach on 5 May 1915. His role as QM ensured he spent most of his time at Helles at the battalion dump at W Beach. But on 7 August, on hearing that most of the battalion's officers had been killed or wounded, Captain William Henry Griffiths, DCM, 'South African War veteran, cool, caustic and clever,' left his base and reached the firing-line at a time of extreme crisis. As Watkins recounts with great poignancy, Griffiths died leading a counter-attack in the Vineyard on 7 August 1915, age 55. Griffiths has no known grave and is commemorated on the Helles Memorial, panel 59.

Griffiths accumulated many medals and awards during his army career: the Bronze (Khedive) Star, the Queen's South African Medal with three clasps, the Kings South African Medal with two clasps, a Mention in Despatches and the Distinguished Conduct Medal.

He was a born leader of men, who by personal example, roused 'thoroughly demoralised' Fusiliers, who followed him as he limped towards certain death.

As with so many other men at Gallipoli, his final gallant actions were unseen by officers, and his exceptional bravery passed unreported and unrewarded. It is only thanks to the recollections of Watkins that evidence survives of Griffiths' last heroic intervention and how his leadership that day turned defeat to victory.

Captain W.H. Griffiths.

'Bill Howarth'

CHAPTER 28

We sat around, quiet, for a bit — the blokes staring down glumly at their boots, scraping bits of mud off them, or twiddling with the tapes of their puttees — an awkward embarrassing spate of silence, every man deep-wrapped in his own private musings. It was left to Bill Howarth, one of our quieter types, to break the silence. In a low voice he asked, 'Sarge, does ta believe there's a God?'

Startled, the Sergeant regarded the questioner curiously. 'That's a bloody funny question to be askin', Bill. Why, doesn't ta believe in Him thiself, then?'

'Ah'm askin' thee, Sarge,' said Bill pointedly.

'But why ask me? Ask t'other blokes. Why pick on me, mate?'

But Bill with an impudent grin said, 'Well, tha's a Sergeant. Tha ought to know better'n me. Ah'm nobbut a bloody private.'

Probably Pte 8593 W. Howarth, who arrived on the peninsula on 5 May 1915. He survived the war and was disembodied on 23 February 1919.

APPENDIX III

Notes on events and places mentioned in the text

'The notorious Krithia vineyard' Battle of 6/7 August at Helles

CHAPTERS 2 AND 23

… I waited till nightfall and eventually stumbled on my company, or what remained of it. There were some grim stories I was to hear. It was all too evident our local push had been a disaster. In the comparative quiet of the next few days, I was to see in various spots of this ragged terrain the price paid for glory. Bodies lay about in clusters, in pairs, singly — and at all sorts of unexpected places where some of the wounded had dragged themselves off to die. And in a corner of that little Krithia vineyard, it was there a corporal and I came across the 6th Battalion, Manchester Regiment boys I have mentioned earlier.

The Battle of the Vineyard was an intense struggle, carried out over a piece of ground little bigger than a football pitch. It began on 7 August and went on into mid-August as part of what was officially designated 'The Battle of 6/7 August at Helles.' Although the troops who took part in the battle had been told their objective was to capture Krithia and Achi Baba, the true objective was to deflect enemy attention from the landings at Suvla and the planned attacks at Anzac.

On 6 August 1915, the 88th Brigade and 1/5th Manchesters made an attempt to advance across Fir Tree Spur, which ended in failure. The next day the 125th and 127th Brigades advanced across Krithia Spur. Both brigades reached their objectives but by mid-afternoon the 127th Brigade and half the 125th Brigade were back in their original firing-line. Only the 1/6th and 1/7th Lancs Fusiliers remained forward, stubbornly holding onto a mere 150 yards or so of enemy firing-line that ran along the north-east end of the Vineyard *(see maps 6 and 7)*.

This tiny piece of ground, won at the cost of many lives and hundreds of casualties, represented the net gain of two days of bitter fighting and its retention took on a totemic significance.

At the height of the fighting, two entire battalions, the 1/6th and 1/7th Lancs Fusiliers, together with troops from all four battalions of the 126th Brigade were involved in the Vineyard's defence. Between them, they fought off repeated counter-attacks made by a determined and courageous enemy.

Arguably the most significant of the reinforcements were the men of A Company, 1/9th Manchesters, led by Captain William Forshaw and 2nd Lieut. Charles Earsham Cooke. For 43 hours Forshaw and his men repulsed the almost continual attacks made against the north-west corner of the Vineyard. The division's GOC later wrote, 'Had it not been for his [Forshaw's] personal example and gallantry, it is questionable whether this corner would have remained in our hands, and its loss would as likely as not have led to the abandonment of the Vineyard as a whole.'[43] Forshaw received the Victoria Cross for his actions, Cooke the Military Cross and three of Forshaw's NCOs the Distinguished Conduct Medal.

The 1/6th Lancs Fusiliers played a vital role in the Vineyard's capture and defence and this is reflected in the awards it received. Four of the battalion's officers and two of its other ranks received Mentioned in Despatches, Pte John Cryer was awarded the Distinguished Conduct Medal, and the CO, Major Roderick Livingstone Lees, VD, who was in overall charge of the defence of the Vineyard, the Distinguished Service Order. Unsurprisingly, the battalion suffered heavy casualties, in fact,

---

43 Report on Operations, August 8th to 31st 1915, by Major General William Douglas, GOC 42nd (East Lancashire) Division, in the 42nd Division, General Staff war diary.

the three days the 1/6th Lancs Fusiliers spent fighting in the Vineyard proved the costliest of its entire service at Gallipoli.[44]

Watkins tells how he came across men of the 6th Manchesters, 'dozens of these lads,' lying head to foot in a corner of the 'notorious Krithia vineyard,' their bodies packed close together, 'festering and swollen in the hot sun.'

The 6th and 8th Manchesters advanced as a combined battalion (both were less than half strength) immediately on the left of the 125th (Lancs Fusiliers) Brigade — from the East Krithia Nullah to the brigade boundary (G11), about 100 yards from the western edge of the Vineyard. Although units inevitably crossed boundaries, it is unlikely that so many of the 6th were killed in the Vineyard proper. Watkins may be using the Vineyard to describe the general area.

'The bit of trench we won with surprisingly light casualties'

CHAPTER 11

Although we'd won that bit of trench fairly easily, we had to struggle hard the next two weeks to hold it, for Johnny Turk is a doughty fighter. Many times the rifle barrel and stock got so hot to the touch that we had to hold it with a bit of thick cloth to avoid burning our hands, and the ground in front of us became littered with Turkish dead. By the end of these two weeks we were about all-in. It had been an endless task in the daytime and at night a continuous shift of sentry duty on the fire-step — one hour on, one hour off, in turn, right through the night. Ask any soldier who's done a couple of weeks like that — ask him what it's like — but before he answers and if you're a bit squeamish on obscenity, first stuff your ears with cotton wool. So, when we stumbled along the gully one night a fortnight later on our way to the beach on being relieved by fresh troops, we were exhausted. In the picturesque littoral of the cotton towns, we were 'bloody well on our knees.' On our way down we passed the troops relieving us, New Zealanders just landed that day, full of starch, self-confidence, brash,

---

44 See Appendix VIII, Establishment, Drafts and Battle Casualties.

bronzed and healthy — not like us, wan and forlorn. One of our chaps called out as we passed, 'Give 'em hell, lads, show 'em what you can do.' They called back, cocksure and confident, 'Just wait till we get at 'em; nobody knows what we'll give 'em.'

We relieved these New Zealanders some ten days later.

Chastened and quiet, they passed us glumly without a word on their way back to the beach for a rest.

Watkins' memory has clearly extended the timescales relating to the events he describes in this passage. The 1/6th Lancs Fusiliers moved into the firing-line running across Gully Spur on 5 May and were relieved by 1st KOSBs and the 2nd South Wales Borderers on the evening of 7 May. The battalion, together with the Lancashire Fusilier Brigade then moved via Gully Beach to a bivouac area above the cliffs south of W Beach. The New Zealand Infantry Brigade arrived at Helles in the early morning of 6 May and, after taking part in the Second Battle of Krithia, moved into Bridge Bivouac on 12 May where it stayed until it embarked for Anzac on 19 May. The 1/6th Lancs Fusiliers and the New Zealanders would have passed each other in Gully Ravine on the evening of 7 May.

The Fusilier Bluff firing-line

CHAPTER 12

The quiet fortnight ended in this trench; we were told to get ready to move that night. It was not to be to the beach for another spell of rest but to occupy another part of the front line abutting off the Western coast line of the Peninsula, a part of the battlefield at present unknown to us...

Watkins is referring to the Fusilier Bluff firing-line that ran down an 'indistinct' spur from Fifth Avenue to the sea. The 1/6th Lancs Fusiliers occupied this firing-line three times between 2 and 18 September. It was the Lancs Fusilier Brigade's only spell in this part of the firing-line; thereafter the Brigade was deployed on the division's Right Sub-section on either side of Border Barricade *(see map 8)*.

The diversionary actions at Helles on 19 December 1915

CHAPTER 16

... two or three days before the evacuation in fact, the Turks were making a last desperate attempt to break through our front line. They were unlucky, but it was fierce while it lasted ...

On 19 December, the 1/6th Lancs Fusiliers were involved in one of three limited diversionary actions at Helles. These were intended to focus the enemy's attention while Suvla and Anzac were being evacuated. One attack was carried out by troops of the 52nd Division at the bifurcation of Krithia Nullah. Another was carried out by troops of 126th Brigade against enemy positions opposite Fifth Avenue at the top of Fusilier Bluff. The third attack was carried out by troops of the 125th Brigade against an enemy position known as the Gridiron opposite Birdcage Walk, immediately east of Border Barricade *(see map 8)*. The 1/7th Lancs Fusiliers carried out the attack supported by 1/6th Lancs Fusiliers who provided the firing-line garrison. The attack was initiated by firing a mine which extended the existing Western Crater and exposed two ends of the enemy's firing-line. The new crater was immediately occupied by the assaulting troops and work began on consolidating the crater. The enemy counter attacked around 21:45 and reoccupied the crater and their firing-line. Capt. Boyd, 1/7th Lancs Fusiliers, organised a counter-attack and succeeded in taking back the crater at 22:00. The following day the new crater was renamed Boyd's Crater 'in recognition of the good work done by Captain Boyd.'

The great storm

CHAPTER 26

... the Cosmic bombardment opened up with a continuous roar of thunder so loud you'd think it would blast the whole world apart. The lightning flashed and played in one continuous whole. The sound even of our own

big guns' bombardments was dwarfed into puny insignificance compared with what came out of the skies, and as if realising the futility of this competition in noise, all the guns on both sides became silent and gave up. Then the sluice gates of Heaven opened and the cold rain fell drenchingly, not in drops but literally in buckets. In less than a minute the floor of the trench was a sea of mud, ankle deep. Ankle deep mud was not strange to us but it was with real alarm we saw the trench itself filling fast with water. Soon it was knee deep and we climbed on to the firing step to avoid it. But it still rose higher and higher until the firing step itself disappeared from view. Little tucks in the side of the ground in the front of our trench became deep crevices under the weight of the falling rain, and the water started to stream in through them, in countless little rivers. By now the water was a good three feet high and rising fast as it was fed still more by the water swirling down from the trenches on the higher ground.

The storm hit Helles on the afternoon of 25 November, with heavy rain quickly turning to snow. Four bitterly cold days of rain, sleet, snow and frost followed, and although conditions for the troops in the firing-line were extremely harsh there are no records to support Watkins' story of flooded trenches and informal truces. However, similar events did occur during the storm at Suvla and it must be assumed Watkins' is relating another veteran's recollections which have become merged with his own memories.

Christmas Day

CHAPTER 28

Then all of a sudden and before you're even aware of it, Christmas Day looms up. On the morning of the sacred 25th, you waken to find the whole war-scarred countryside has been blanketed overnight by a heavy fall of snow. From our high vantage point, we gaze down across the main gully. The ragged terrain has been smoothed off considerably by the soft thick blanket of snow and the snow sparkles in the morning sunlight. It's beautiful to look at and you feel a lifting of the heart as you gaze over the countryside. A stainless white blanket — hiding nearly all the ravages of war, stainless — except for the pink-splashed snow near your trench from the blood of the wounded as they had staggered by after last night's hand-grenade attack...

Down below in the gully, the apotheosis of incongruity. Incredibly, on this Devil's own playground, someone had unearthed the Battalion's band instruments from their temporary storage and the bandsmen were gathered down there in the gully, their polished instruments winking and flashing in the morning's bright sun, and the strains of Christmas carols were wafted on the still morning air...

Watkins' recollections of Christmas Day are in fact an amalgam of memories of at least two separate occasions. It is well known that it did not snow on Christmas Eve or Christmas Day at Helles, and the only significant falls of snow at Helles were on the nights of the 28/29th and 29/30th of November. At that time the 1/6th Lancs Fusiliers were occupying that section of the Eski Line that ran across Gully Spur, which was close to, if not the same, line of trenches taken by the battalion on 6 May 1915. The Eski Line formed the last line of defence for the beach heads and as such was wired and constructed as a 'fighting trench' with a fire step. A battlefield cemetery known as H15 was situated in Gully Ravine close to where the Eski Line crosses the gully and this may well have been the cemetery Watkins mentions in his account. On Christmas Eve, the 1/6th Lancs Fusiliers were relieved from the Birdcage Walk firing-line on the right of Border Barricade. The battalion moved back into winter quarters at 'Sites 6 & 7, Gully Ravine,' while the machine-gun sections moved into the Eski Line West. Watkins could have listened to the battalion band playing in the gully on Christmas Day from either of these positions.

The evacuation

CHAPTERS 29 AND 30

In the first week of January, ranks began to thin out, blokes departing silently to appointed rendez-vous on the beach. I never actually knew what happened or how it was organized, but chaps would be there with you one day in the trenches and the next day they weren't there. Came the last night and a call for volunteers — just a few — to man the sectors of our front line trench until the last possible minute while the remaining body of troops got away and to give the Turks a deceptive semblance of normal occupation and life — keeping up our usual nightly intermittent rifle fire.

I volunteered…

Watkins describes being among the last to leave the front-line trenches on the night of 8/9 January 1916 when the British Army completed its evacuation from Helles. However, his battalion was taken off the peninsula on 27 December 1915.

The 125th Brigade war diary, describing their evacuation, notes '100 employed men remaining on the Peninsula till relieved.' But it does not indicate when these men were embarked. They were probably a beach party dealing with the Brigade's baggage, not dispersed into another division's line.

But some men of the 42nd Division did indeed remain on the peninsula until the final night, as the divisional history reveals:

> The Divisional Artillery remained behind, and also a small detachment of Engineers and the 1st and 3rd Field Ambulances, all attached for duty to the 13th Division […] Some of the gunners and the greater part of the R.A.M.C. left a few days before the curtain fell on the final scene of the great tragedy of Gallipoli. The last men of the 42nd Division — and among the very last of the allied forces — to leave the peninsula were detachments of artillery and R.A.M.C. and a small party of Engineers…

> Late at night the troops began to leave the firing-line. When they had passed, the men in the second line filed out, and after them followed the small parties of — each of one officer and four men — of the East Lancashire R.A.M.C. to pick up stragglers and assist any sick or injured. Last of all came the handful of sappers who had charge of closing the gaps in the entanglements of Gully Ravine.[45]

---

45 Frederick P. Gibbon, *The 42nd (East Lancashire) Division: 1914–1918*, pp. 60, 62.

APPENDIX IV

42nd (East Lancashire) Division, Order of Battle & Field State, 2–5 May 1915

Showing all units that embarked for Gallipoli

Details of personnel and appointments are taken from the 42nd Division War Diary — Appendix 37 — Order of Battle 29 May 1915, with the exception of the CO of the 1/9th Manchesters, Lieutenant Colonel D. H. Wade, who had been wounded and evacuated by this date.

Divisional Headquarters

GOC, Major General William Douglas, CB, DSO
Captain H. T. Cawley, MP, ADC
Lieutenant J. W. Fry, ADC
Lieutenant Colonel A. W. Tufnell, GSO 1
Lieutenant Colonel F. A. Earle, GSO 2
Captain S. H. Kershaw, GSO 3
Colonel E. S. Herbert, AA & QMG
Captain R. S. Allen, DAA & QMG
Major R. J. Slaughter, DAQMG
Colonel J. Bentley-Mann, ADMS
Captain C. M. Drew, DADMS
Captain Briercliffe, Sanitary Officer

Lieutenant Colonel T. Marriott, ADVS
Captain J. Magill, VO
Major O. R. E. Milman, DADOS
Captain T. B. Forwood, APM

Divisional Signal Company

Major A. W. Lawford

Divisional Artillery

Due to the lack of space for artillery at Helles, only the 5th Lancs Battery and two guns of the 6th Lancs Battery were landed, all other batteries and guns being returned to Egypt.

1/1st East Lancs Brigade RFA

    4th Lancs Battery

    5th Lancs Battery (grouped with 29th Div. Artillery)
        Major J. C. Browning

    6th Lancs Battery

    1st Ammunition Column

    2nd Ammunition Column

1/3rd East Lancs Brigade RFA

    18th Lancs Battery

    19th Lancs Battery

    20th Lancs Battery

    3rd Ammunition Column

Divisional Engineers

Lieutenant Colonel S. L. Tenant, CRE

    1/1st East Lancs Field Company
        Major J. H. Mousley

    1/2nd East Lancs Field Company
        Major L. F. Wells

125th (Lancashire Fusiliers) Infantry Brigade

The main recruiting centre for each of the division's infantry battalions is shown in brackets.

Brigadier General H.C. Frith, CB
Brevet Major A.J. Allardyce, Brigade Major
Captain J.C. Kenyon, Staff Captain

    1/5th Bn Lancashire Fusiliers (Bury)
        Lieutenant Colonel J. Isherwood, VD

    1/6th Bn Lancashire Fusiliers (Rochdale)
        Lieutenant Colonel Lord Rochdale

    1/7th Bn Lancashire Fusiliers (Salford)
        Lieutenant Colonel A.F. Maclure, TD

    1/8th Bn Lancashire Fusiliers (Salford)
        Lieutenant Colonel J.A. Fallows, TD

126th (East Lancashire) Infantry Brigade

Brigadier General D.G. Prendergast, CMG
Major C.J. Hickie, Brigade Major
Captain T.C. Robinson, Staff Captain

    1/4th Bn East Lancs Regt (Blackburn)
        Lieutenant Colonel F.D. Robinson, VD

    1/5th Bn East Lancs Regt (Burnley)
        Lieutenant Colonel W.E. Sharples, TD

    1/9th Bn Manchester Regt (Ashton under Lyne)
        Lieutenant Colonel D.H. Wade

    1/10th Bn Manchester Regt (Oldham)
        Lieutenant Colonel J.B. Rye, VD

127th (Manchester) Infantry Brigade

Brigadier General Noel Lee, VD
Major H. L. Knight, Brigade Major
Captain T. C. Nevill, Staff Captain

> 1/5th Bn Manchester Regt (Wigan)
> Lieutenant Colonel H. C. Darlington
>
> 1/6th Bn Manchester Regt (Stretford, Manchester)
> Major C. R. Pilkington
>
> 1/7th Bn Manchester Regt (Manchester city centre)
> Lieutenant Colonel H. E. Gresham, TD
>
> 1/8th Bn Manchester Regt (Ardwick, Manchester)
> Lieutenant Colonel W. G. Heys, TD

East Lancs Divisional Train ASC

Due to lack of space at Helles only No. 2 Company was allowed to land with the division. The other three companies were then sent back to Egypt.

> No. 1 (Headquarters) Company
> No. 2 (Lancs Fusiliers) Company
> Major A. England
> No. 3 (East Lancs) Company
> No. 4 (Manchester) Company

Divisional Royal Army Medical Corps

Although the 1/1st and the 1/3rd field ambulances both landed with the division, only C Section and part of B Section of the 1/2nd Field Ambulance landed at Helles on 10 May, with the remainder landing on 17 June 1915.

> 1/1st Field Ambulance
> Lieutenant Colonel H. G. Parker
>
> 1/2nd Field Ambulance
>
> 1/3rd Field Ambulance
> Lieutenant Colonel W. M. Steinthal

Summary of arms on embarkation for Gallipoli

	Officers	Other Ranks	Guns	MGs
Headquarters and Signal Company	25	249		
Artillery	57	1,257	24	
Engineers	15	394		
Infantry	386	10,830		24
ASC	24	313		
RAMC	30	644		
Total	537	13,687	24	24

Summary of arms for 15 September 1915

	Officers	Other Ranks	Guns	MGs
Headquarters and Signal Company	19	218		
Artillery	23	294	18	
Engineers	21	341		
Infantry	160	5,302		24
ASC	10	126		
RAMC	24	481		
Total	257	6,762	18	24

The loss in the division's field strength is particularly significant given the figures above include infantry reinforcements of approximately 100 officers and 3,500 other ranks. The 18th Battery RFA and the 1/2nd West Lancashire Field Company RE had also joined the division in the interim period.

APPENDIX V

Excerpt from

'The Lancashire Fighting Territorials' *by* George Bigwood (1916)

6th Battalion Lancashire Fusiliers (Rochdale)

George Bigwood was a Manchester-based journalist who went on to become a military correspondent and author of books on the cotton trade. His best-known work however is The Lancashire Fighting Territorials *which he produced a few months after the campaign ended. Despite the book's short gestation, it is a largely reliable account that relies heavily on the personal accounts of those who served in the campaign, many of which were reproduced in national and local newspapers. Bigwood also kept up a regular correspondence with General Douglas, the division's commander, both during the division's time in Egypt and when it fought at Gallipoli. Although General Douglas would have presented the official line, it does not seem to have skewed the accuracy of Bigwood's description of events and may well have assisted him in forming a broader picture.*

Lord Rochdale's men landed on the Gallipoli Peninsula a few hours before their Bury comrades [the 1/5th Battalion Lancs Fusiliers]. This Battalion sailed from Egypt on May 1, 1915, reached the theatre of operations on May 5, and on the evening of that day were in the trenches before Krithia, once a collection of whitewashed houses, now a heap of ruins with the dominating hill Achi Baba rising immediately in its rear. The passage from Egypt to the Dardanelles was uneventful, still there was plenty of life to interest the Lancashire troops who had spent six months on the Egyptian desert. The seriousness of their mission was brought home to them with full force when they approached the place of landing. For a long time they had heard the distant boom-boom of guns; they were now near enough fully to appreciate what is meant by naval supremacy. The guns were trained on the Achi Baba position and the Turks replied, but ineffectually, for all their shells dropped harmlessly into the sea. Some of them, it is true, got perilously near our ships, but happily not near enough to do any damage.

'We landed,' writes one of the Fusiliers, 'during a heavy and continuous bombardment. We reached the battlefield under the cover of darkness. As long as I live I shall not forget the experience. Our line of march was up the cliff and over some hills and undulating ground to the trenches. We each carried two hundred rounds of ball. With the break of day we received the order to advance and were met with a shower of shrapnel. We could not all live in these conditions, so that you will not be surprised to hear that the field was quickly covered with casualties. Some were killed; many were wounded, but we had not time to attend to any of them, for we were engaged in a life-and-death struggle and it might easily be our turn to fall by the wayside. How I got to the trenches without being hit I do not know. It was a miracle that any of us live to tell the tale. Captain Scott was badly wounded. The Turks, though bold and courageous fighters, do not like our bayonet charge. When we closed in on them they would throw up their hands and shout invocations to Allah, but we did not withhold the bayonet on this account.'

Major W. D. Heywood also praises the fighting spirit of the Rochdale men. Writing home in the early days of the campaign he says:

'We are proud that the Sixth were the first battalion to get to the firing line, and I think it a tribute to the men that, taken as they were from peace conditions in Cairo and thrown straight in the firing line, they behaved so splendidly. They never hesitated to advance in the face of the machine gun and shrapnel. In the first day's fighting all the officers were splendid. Lord Rochdale and Major R. L. Lees and Captain Spafford (adjutant) bore charmed lives. They walked about as if on a parade ground. Captain Scott (wounded) led B Company with great dash.'

Major Heywood said he would not forget in a hurry their first night in the trenches. It was an inky black night, and owing to losing their way they had to sit in a gully for two hours. Had the Turks known their predicament they could have wiped out the whole detachment with shrapnel. The bullets from the firing line were whistling over their heads. At last they got into a trench about 400 yards in rear of the firing line. The scene here was extraordinarily weird. Shells from both sides burst with bright flashes, and the coloured flare lights and the roar of the guns combined to make a fascinating, though terrible picture.

When the Lancashire Fusiliers Brigade joined the Mediterranean Expeditionary Force our troops had forced their way forward some 5,000 yards from the landing-place at the point of the Peninsula. Opposite them lay the Turks, who, since their last repulse, had fallen back about half a mile upon previously prepared redoubts and entrenchments. The situation when the Territorials entered the fighting area is best described in Sir Ian Hamilton's own words:

'Both sides had drawn heavily upon their stock of energy and munitions, but it seemed clear that whichever could first summon up spirit to make another push must secure at least several hundreds of yards of the debatable ground between the two fronts, and several hundred yards, whatever it might mean to the enemy, was a matter of life or death to a force crowded together under gun fire on so narrow a tongue of land.'

Before this date (May 5) progress had been slow, and the fighting of a decidedly critical nature. We had made good our landing against a numerically superior force, and the fringe of country in our possession as well as the general character of that country made a forward rush imperatively necessary. Freedom of manoeuvre was impossible, and many of the ordinary rules of warfare had to be set aside and the initiative of the General Officer Commanding relied upon to get over the extraordinary, the unprecedented difficulties in which our troops found themselves. The ground held was exposed to the incessant bombardment of well-placed guns on Achi Baba, and all the supplies both as regards ammunition and food had to come overseas. Troops usually depend upon the country they hold to provide a great deal of their subsistence.

But this country was absolutely barren — a veritable wilderness. 'The country is broken, mountainous, arid, and void of supplies,' was the description Sir Ian Hamilton gave of it. The General adds:

'The water found in the areas occupied by our forces is quite inadequate for their needs; the only practicable beaches are small, cramped breaks in impracticable lines of cliff; with the wind in certain quarters no sort of landing is possible; the wastage, by bombardment and wreckage, of lighters and small craft has led to crisis after crisis in our carrying capacity; whilst over every single beach plays fitfully throughout each day a devastating shell fire at medium ranges.'

This explains why the Rochdale and other battalions of the Lancashire Fusiliers were thrown into the fight as soon as they set foot on the Peninsula. There was no alternative to such a course of action, and as the other battalions of the brigade landed they, too, were marched against Krithia. On May 6 Lord Rochdale received orders with other troops vigorously to attack, for it was decided that ground must be gained at all costs. The men had to begin the engagement at eleven o'clock. Waiting for that hour was really worse than charging with the bayonet. There is little excitement crouching in a trench waiting for the impending fight, and it is not a good thing for the nerves. Presently the fateful hour struck and the dash was made. 'Every yard was stubbornly contested; some brigades were able to advance; others could do no more than maintain themselves. Positions were carried and held, other positions were carried and lost.' The gunners lengthened the fuses of their shrapnel, and their effective fire compelled the enemy to yield ground which our brave Rochdale men and their comrades took and held. For five hours the battle raged. Big guns, small guns, musketry fire, the bayonet, and the bomb were used in the contest, and both sides suffered heavily in killed and wounded. Hidden machine guns swept the Rochdale position and caused much havoc, but the men clung tenaciously to what they had won and were courageously facing the withering fire in order further to break down the enemy's power of resistance. But the French Corps had found a terribly hot corner — the enemy opposed to this Corps was well entrenched behind an earthwork, and able to offer a severe check to our ally, who could not dig themselves in until night came on and stopped the fight.

The Lancashire Fusiliers began the attack on the following morning at ten o'clock, but the hidden German machine guns and the Turkish snipers hidden among a clump of trees made it impossible to advance without a tremendous sacrifice of life. The troops had been fighting continuously for five hours and were now exhausted, but they were not too exhausted successfully to assault the Turkish position at the point of the bayonet. They gained some 300 yards of ground and occupied the first line of Turkish trenches, which contained many dead bodies. At sundown new trenches were dug and the attack renewed the following morning. The Fusiliers were now withdrawn into reserve.

These initial engagements are tersely described by the soldiers as 'hell.' They claim that there is no other word known to them which so nearly applies to the conditions under which they fought. The Rochdale men had repeatedly to rush over an open piece of ground when the enemy's machine guns were literally sweeping the whole fine with bullets (one of these guns can fire up to 400 rounds a minute, and the volume of fire is equal to the rapid fire of thirty men and very deadly), and the effective 'bursts' of shrapnel were pouring showers of lead among them. But they never flinched. Every man of them went through this terrible ordeal with that bravery which will not accept defeat. They often essayed the impossible, and the death-dealing missiles thinned their ranks as they suddenly ceased their fire and rushed at the enemy with the bayonet. The word 'advance' seemed to bring new life to the battalion. Men in various stages of exhaustion stumbled over the rough ground and fell upon the Turks with a spirit and dash which would have done credit to any of the troops on the Peninsula. Lord Rochdale had every reason to be proud of their gallantry.

'Bullets were whistling all around us and bursting shells added to the murderous fire,' wrote one of the Fusiliers. 'Our officers are a credit to England, especially Lord Rochdale, Captain Scott, and the Adjutant. There was Lord Rochdale giving his orders and saying, 'Give it 'em, lads!' and we were doing it for all we were worth. When night came and everything was quiet. Lord Rochdale said: 'Lads, I am proud of you. You have captured the hill; you have done twice as much as we expected you to do!'

Among the officers wounded were Captain J. J. Gledhill, Captain G. Scott, Captain R. W. Leach, Lieutenant L. Maurice Robinson, Lieutenant P. V. Davies, Lieutenant J. S. Berrington, Lieutenant M. C. De Wiart, Lieutenant J. S. Lord, Lieutenant T. D. M. Bartley, Lieutenant Eric Molesworth, Lieutenants J. and W. Taylor, and Lieutenant W. Redmond.

'We had an awful time,' Captain Gledhill wrote from a hospital in Malta. 'When D Company left the advanced trenches we met the full blast of the Turkish maxims, shells, and rifle fire. It was just like a hailstorm. Our men began to drop, but the rest pushed on. They did splendidly. They were followed by C Company (Captain R. Barker, of Todmorden, in charge), so you see the Todmorden lads were in the van. We left the trenches at 11.30, and I was hit at 12.15, concurrently by a machine-gun and a shrapnel bullet. My right arm was broken and badly shattered... Here I am absolutely in the dark as regards the casualties among my men. I have seen no casualty lists, and I only hear fragments from some of my men who come to see me... I learn that my company has been hard hit, there being only one officer left out of six.'

Subsequently the Fusiliers had to repulse some severe counter-attacks. Occasionally the Turks effected a temporary lodgement and they were driven out by the bayonet. Every night up to June 3, assaults were made on the redoubt, and upon our line of skirmishers, but at the end our position remained intact. On June 4, the day of the big attack by the Manchester Brigade, and subsequent days the Rochdale Fusiliers were again in fierce contact with the enemy. The 6th Lancashire Fusiliers acted in conjunction with the Manchester Brigade, and took a prominent part in the capture and maintenance of the Turkish trenches.

After a fortnight's rest on the Isle of Imbros, they were back in the trenches on July 10, being in action for a third time a week later.

The worst experience of the 6th Fusiliers came on August 7, when they took part in a new offensive, which had begun the previous day with a fresh landing of troops at Suvla Bay. With others of the Lancashire Brigade the Rochdale men were ordered to engage the enemy's attention by an attack on the Achi Baba position in the south, while the new landing was pushed forward.

Of the fifteen officers of the 6th Battalion who went over the parapet to attack, eight were killed, six were wounded, and only one came back unhurt. The result of this attack was the capture of the vineyard by the 6th Lancashire Fusiliers, and despite many counter Turkish attacks, they held the ground till they were relieved some days after. This was the only ground that was taken and held at the Helles end of the Peninsula on this occasion.

Sir Ian Hamilton in the dispatch which covered these operations said:

'Both our Brigades had lost heavily during the advance and in repelling the fierce onslaughts of the enemy, but, owing to the fine endurance of the 6th and 7th Battalions of the Lancashire Fusiliers, it was found possible to hold the vineyard through the night, and a massive column of the enemy which strove to overwhelm their thinned ranks was shattered to pieces in the attempt.'

Later, life in the trenches was not quite so strenuous, but the hardships of the campaign had not lessened. Bombardments were continued daily, and sniping and trench work kept the men busy. Amid all the deafening roar of modern battle some attention is paid to the spiritual welfare of the troops. The Rev. Denis Fletcher, the chaplain attached to the Rochdale unit, wrote home to say that it was impossible to held church services.

To gather a big crowd of men in the open air was to invite attention from the enemy's guns. At one of his celebrations of Holy Communion his altar was two wooden boxes of provisions. The men knelt all round and the service was most impressive. 'The men have to learn under these conditions that religion is a thing in themselves and that they must not depend on church or chaplain.'

Lord Rochdale commanded the Lancashire Fusiliers Brigade in the engagement of May 6 and 7, and temporarily succeeded Brigadier-General Noel Lee in command of the Manchester Brigade. He was invalided home in October with typhoid fever.

On Christmas Eve the battalion left the firing line after a week full of excitement and danger. One of the enemy's trenches had been captured. During this exploit the Rochdales again showed the Turks that they were still full of fight, and that the bayonet was a weapon which they could use skilfully and with effect. That night they enjoyed the luxury of sleeping in hutments which were to be the winter quarters. Christmas Day was spent happily, though quietly. The rations marked it off from the ordinary day. But there were many vacant places; familiar faces were missing. It was known that some were resting in the new graves in the gully. They had made the supreme sacrifice only that week. Others had fallen in the fight and had been lost to them altogether. Christmas, 1915, therefore, was a time of chastened gaiety. Still, it is not well that the soldier on the battlefield should dwell on the past. Soldiers with drooping spirits cannot fight. It is right, it is fitting, it is indeed one of the characteristics of a great soldiery not to be insensible to the loss of gallant comrades. On the other hand it is fatal to success in the field to allow a feeling of depression to pervade the ranks. The Rochdale Fusiliers guarded against this. As the shadows of departing day crept over the stricken field there was a rattle of musketry. This stirred the men out of the reverie which a day out of the trenches had encouraged. The trench that they had captured in the previous evening was being attacked, and the battalion had orders to prepare to contest the right of ownership when the firing became of a more desultory character and gradually died away. Presently a messenger brought the news that the trench had been vigorously attacked by the enemy and successfully defended by the 7th Manchesters.

There are many days in the life of our men on the Peninsula that will never be forgotten as long as they live. The Rochdale soldiers will always be proud of the fact that they were the first of the Divisional units to enter the fight. Boxing Day, 1915, will not be forgotten, for it was on the night of that day that the news of their early departure reached them. 'The Battalion must be prepared to embark to-morrow,' was passed from hutment to hutment and from man to man.

'The morning came, and the news held good,' wrote one of the Sixth. 'The Battalion was to move in the afternoon in full marching order. The day wore on, and before the time came the men 'fell in' loaded up with equipment, oil-sheet, blankets, etc., and two days' rations. On the order to move, the first company filed into the gully and immediately went up to the knees in mud. It is not far from Eski Line to Gully Beach, but it took nearly two hours before the latter place was reached, such was the difficulty of negotiating the mud-choked gully. The 'dump' looked strangely deserted, and that goal of the ambitious, the divisional canteen tent, was nodded a last farewell.

'The sun had set as we took the road along the shore. Imbros, a dark mass, was sharply defined across the water, while Samothrace, farther north, was bathed in the rapidly darkening afterglow. A few hundred yards were tramped, a halt was called, and the men, already somewhat weary with their heavy load, were glad to rest. Darkness fell, and the night grew chilly and cold, but no move was made for several hours. At ten o'clock we started again in file along the beach, picking our way gingerly over cobbles and wet sand. Soon we ascended the cliff and found a dull, smoky moon rising behind Achi Baba. Away back from the distant left came the rattle of the firing line. On our right along the top of the cliffs were the remains of the trenches from which a murderous fire had been poured at the first landing. Winding downwards, the base was reached at 'Lancashire Landing' and another halt was called. Away over the eastward a flash was seen. A short pause; then, with a roar and a shriek, a shell hurtled overhead and struck the hillside just behind. A tremendous explosion followed, and showers of metal flew in all directions. It was 'Asiatic Annie' firing 8-in. shells from the far side of the Straits, a distance of five miles. At first it was hoped that it was an odd shot. But no; every ten minutes the flash was seen, and after about half a minute's interval the shell shrieked over. There was not much cover, but every man took what little there was and hoped for the best.

'This rather trying ordeal lasted until two o'clock, fortunately without accident, when word was passed along to file on to the lighter. From the lighter the steamer could be seen looming up in the dark a short distance out. The journey took but a few minutes. Officers and men soon settled down to sleep, and only one or two were aware of a shell that dropped half an hour later into the water a few yards from the vessel, and sending a column of water over the deck.

'When the sun rose the vessel was well out to sea. Away over the stern, bathed in a wonderful morning pink, was the peninsula, the place of high hopes, of disappointments, of sorrows and hardships endured, of danger encountered and duty done.'

Soldiers of the Lancashire Fusiliers on their way to the beaches at Helles on 5 May 1915. A trawler ferrying the men passes HMS *Implacable*.

APPENDIX VI

Timeline, 1/6th Lancashire Fusiliers at Gallipoli, May to December 1915

Timings of events are taken from the Lancs Fusiliers (125th) Brigade war diary and the 29th Division HQ war diary. Passages quoted are from the brigade war diary.

2 May   The Lancashire Fusiliers (125th) Brigade embarks on the following transports at Alexandria.

SS *Menominee*: Brigade HQ, No. 2 Signal Section, 1/5th LFs, part of No. 2 Company ASC, 18 GS wagons and the transport animals for 1/6th & 1/7th LFs.
SS *Nile*: 1/6th & 1/7th LFs.
SS *Karoa*: 1/8th LFs, part of No. 2 Company ASC, 18 GS wagons and a 'few' horses.
In addition, 43 carts 'for use as 1st line transport' were loaded on SS *Uganda*.

The Lancashire Fusiliers Brigade was redesignated the 125th Infantry Brigade on 25 May 1915. However, for simplicity this latter title will be used in its abbreviated form throughout this appendix.

4 May   SS *Nile* arrives off Cape Helles at 18:00.

273

5 May	1/6th LFs disembark through the SS *River Clyde* at V Beach at 12:30. Shortly afterwards, battalion moves to bivouac area above W Beach.
	During the night the battalion moves up Gully Ravine and takes over the firing-line across Gully Spur, relieving the 1st Kings Own Scottish Borderers (KOSBs).

Second Battle of Krithia

6 May	
05:00	125th Bde receives orders 'for attack on enemy's position about Achi Baba.' The 1/6th LFs ordered to capture enemy trenches to their front.
11:00	88th Brigade (29th Division) advance.
11:54	1/6th LFs 'reported ready to advance from its trenches.'
12 noon	1/6th LFs advance, and are checked by heavy enemy machine-gun fire, causing the battalion to halt a short distance in front of its objective.
14:30	One company of 1/7th LFs sent to reinforce the 1/6th LFs.
15:15	The 1/8th LFs and the remainder of 1/7th LFs were sent forward into the firing and support trenches.
7 May	
06:15	Orders issued by brigade HQ to resume advance at 10:00, however, 'those for 6th & 7th LFs went astray.'
10:00	'Infantry attack commenced… The advance did not succeed in going forward so far as was desired. Enemy machine-guns were particularly active, and were not located. Many casualties were down to them.'
13:55	'OC, 6th LF, reported that it was impossible to get the men to advance in face of the machine-gun fire which knocked out every man who showed himself.'

14:57	'Order received for 6th LF to advance, in cooperation with KOSBs and Inniskilling Fusiliers who were sent to reinforce and clear the ground from which cross fire from machine-guns was being directed.'
17:50	'OC, 6th LF, reported that the 7th LF, had started advancing twice but in each occasion every man was hit. He considered it impossible to get the men out of the trenches.'
18:00	'OC, 6th LF, reported KOSBs working round left flank. Apparently little, if any advance was made as a result of the day's operations except on the extreme left where a slight advance was reported, the same line of trenches as those occupied before, were held.'
18:26	125th Bde received orders they would be relieved during the night and were to move back to the bivouac area above W Beach. The majority of the brigade reached W Beach by 02:30 on 8 May, however, 'Two platoons of 6th LFs could not be relieved, remained in the trenches.'
8 May	125th Bde attached to the Composite Division.
02:30	1/6th LFs at W Beach bivouac area.
19:35	1/6th LFs together with 125th Bde ordered to move to a bivouac area close to where the road from Seddul Bahr to Krithia crosses the Kirte Dere (later known as Bridge Bivouac or Stone Bridge Bivouac).
Midnight	'All battalions settled in bivouac. Dug-outs, previously used by Manchester (127th) Bde, were unoccupied.'

—

9–10 May	1/6th LFs, together with 125th Bde, at Bridge Bivouac.

APPENDIX VI

11 May	Brigade receives orders to move.
20:00	125th Bde moves from Bridge Bivouac to take over the firing-line and reserve trenches across Krithia Spur (between Achi Baba Nullah and Krithia Nullah).
	1/6th LFs move into the No. 1 Australian Line, as brigade reserve.
13 May	1/6th LFs in No. 1 Australian Line.
17:30	1/6th LFs ordered to move back 600 yards to the Eski Line.
20:10	Move commenced.
14 May to 3 June	1/6th LFs in Eski Line as brigade reserve.
21 May	Lord Rochdale relinquishes command of 1/6th LFs and assumes temporary command of 126th Bde.
25 May	Redesignation of Territorial Formations announced.
	1/1st East Lancashire Division to be 42nd (East Lancashire) Division.
	1/1st Lancashire Fusilier Infantry Brigade to be 125th Infantry Brigade.
	1/1st East Lancashire Infantry Brigade to be 126th Infantry Brigade.
	1/1st Manchester Infantry Brigade to be 127th Infantry Brigade.
28 May	1/6th LFs send a company to occupy section of Eski Line west of Krithia Nullah (to the divisional boundary with 29th Division).
30 May to 3 June	1/6th LFs digging communication trench (A Avenue) in preparation for the coming battle.
3 June	1/6th LFs move into Burnley Road and Mercer Road support trenches after dark.

Third Battle of Krithia

Only key events relating to the 1/6th LFs are mentioned.
See maps 4 and 5.

4 June	1/6th LFs occupying division's portion (around 1200 yards) of Burnley Road/Mercer Road, from the divisional boundary on the left about 120 yards west of Krithia Nullah, to Achi Baba Nullah on the right.
01:15	1/6th LFs receive operational orders from brigade HQ outlining the timetable of events for the advance and detailing each battalion's role. The 1/6th LFs detailed to follow both the attacking waves of the 127th Bde, and, together with the division's 1st and 2nd Field Coys RE, help to consolidate the trenches and ground taken.
	1/6th LFs machine-guns to work independently of the battalion and in cooperation with 127th Bde.
07:00	Lord Rochdale returned from commanding the 126th Bde, and resumed command of the 1/6th LFs.
08:00	Bombardment of enemy trenches commences.
12:00	First Advance goes forward, followed closely by two coys 1/6th LFs and two half-coys of RE.
12:05	First Objective line captured.
12:15	Second Advance goes forward, followed closely by two coys of 1/6th LFs and two half-coys of RE.
12:45	Second Objective line captured.
14:26	Lord Rochdale ordered to assume command of the 127th Bde.
19:00	Order given for the troops of the 125th and 127th Brigades to withdraw from Second Objective line and consolidate on the First Objective line.

5 June	1/6th LFs 'merged' in First Objective firing-line, fighting off Turkish counter-attacks.
6 June	1/6th LFs working parties digging support and communication trenches. Balance of battalion in the Australian Lines as part of divisional reserve.
20:26	Lord Rochdale placed in command of divisional reserve (less troops in the Australian Lines).

—

7–10 June	1/6th LFs in Australian Lines.
11–14 June	1/6th LFs take over the Right Sub-section firing-line (from the Krithia Road to Achi Baba Nullah) on the evening of 11 June and remain until relieved by 1/8th LFs on the afternoon of the 14 June.
14–25 June	1/6th LFs in Australian Lines in divisional reserve.
	Battalion embarks for Imbros on the night of the 25/26th.
26 June to 9 July	1/6th LFs at Imbros for rest and recuperation.
	Lord Rochdale granted leave to return to England.
10 July	1/6th LFs returns to peninsula on night 9/10th. Battalion temporarily attached to the 126th Bde, and moves into the support line.
11–16 July	1/6th LFs in firing-line between Krithia Nullah and Krithia Road.

Action of Achi Baba Nullah

12 July	52nd Division and French attack Turkish trenches on division's right.
13 July	52nd Division, French and RND attack Turkish trenches on division's right.

—

16–20 July	1/6th LFs in Redoubt Line.
19 July	1/6th LFs temporarily attached to 127th Bde.
20–21 July	1/6th LFs in firing-line between Krithia Nullah and Krithia Road.
22–30 July	1/6th LFs in reserve dug-outs or trenches in corps reserve.
30 July to 1 August	Two coys 1/6th LFs in No. 1 Australian Line, two coys in reserve dug-outs or trenches in corps reserve.
1–4 August	1/6th LFs in firing-line either side of Krithia Road.
4–5 August	1/6th LFs in No. 1 Australian Line in brigade reserve.
5 August	Battalion takes over an approximately 140-yard-long section of the firing-line — from the junction with trench G11 to a point just left of centre of the Vineyard.

Battle of the Vineyard – Battle of 6/7 August at Helles

See maps 6 and 7.

6 August	1/6th LFs in firing-line left of Krithia Road.
15:50	88th Brigade and the 1/5th Manchester Regt launch attack west of Krithia Nullah.
7 August	125th & 127th Brigades to attack at 09:40.
	1/6th LFs in firing-line left of Krithia Road.
08:10	Allied artillery bombardment commences.
09:40	First Infantry Assault goes forward (two coys 1/6th LFs following):
	127th Bde, from divisional boundary to the left of Krithia Nullah to brigade boundary, trench G11, 100 yards left of the Vineyard.
	125th Bde, from trench G11 to a point 100 yards east of Achi Baba Nullah.

09:55	125th Bde reports First Objective (G12 & F12) taken and held. 127th Bde reached G12a on 1/6th LFs' left but unable to make progress elsewhere.
	Second Infantry Assault goes forward (two coys 1/6th LFs following). Only the 1/6th LFs reach the Second Objective (G13) but were forced to retire to G12 (known later as Vineyard Trench) due to exposed flanks.
11:00	Strong Turkish counter-attack forces both brigades back to their original firing line. Shortly afterwards troops of 125th Bde counter-attack and regain G12 & F12 — this may well be when Captain Griffiths led the 1/6th LFs' counter-attack that Watkins describes in Chapter 24.
13:30	Turkish counter-attack on 125th Bde's line halted by Allied artillery fire.
14:20	Turkish counter-attack repulsed west of Small Nullah (near NE corner of Vineyard).
15:18	Two platoons, 1/9th Manchester Regt, under Captain William Forshaw sent to reinforce 1/6th LFs holding G12, taking up position at the NW corner of the Vineyard where three Turkish held trenches converged.
	Forshaw remained at his post for 43 hours, repelling numerous Turkish counter-attacks. For his actions Forshaw would later be awarded the Victoria Cross. In addition, Forshaw's 2ic, 2nd Lieut. Charles Earsham Cooke, received the MC and three of Forshaw's NCOs the DCM.
16:00	Strong Turkish counter-attack forces the 1/5th and 1/8th LFs back to their original firing-line, leaving both 1/6th LFs' flanks exposed.

19:05	42nd Division General Staff diary entry: 'At the end of these operations, the front line occupied by the division was the same as the original firing-line, except in the vicinity of the Vineyard, where we held the western edge of the Vineyard, G12, N. of the Vineyard, and about 60 yards of F12 to the east.' (Both trenches being held by 1/6th & 1/7th LFs and elements of the 126th Bde.)
8 August	1/6th LFs holding G12/Vineyard Trench.
	During the night of 7/8 August, a working party of 140 men from 1/4th East Lancs Regt and six sappers under an RE subaltern dug a communication trench along the SE edge of the Vineyard, connecting the NE corner of G12 with the original firing-line. By dawn the trench was wide and deep enough to be used.
04:40	Enemy attack on NW corner of the Vineyard repulsed. Attacks on NW corner continued throughout the day, twice driving Forshaw and his men back but each time they succeeded in retaking it.
21:50	Heavy enemy attack against G12 'driven off by the 1/6th Lancs Fus., with the bayonet. An attack was also made about the same time on the NE corner of the Vineyard, which was driven off by the 1/7th Lanc Fus.' Both attacks were preceded by heavy enemy bombardments.
Night 8/9	Enemy attacks continued throughout the night and a section of the new communication trench on the eastern side of the Vineyard was lost.
9 August	1/6th LFs holding G12/Vineyard Trench.
	Enemy continued to bomb the NW corner of the Vineyard after dawn, rendering the position 'untenable more than once.'

07:00	The enemy finally driven out of the new communication trench.
09:00	The enemy bombed out of the NW corner.
10:00	Situation 'less critical and enemy less active.'
14:45	The 1/4th East Lancs Regt relieved the Vineyard garrison.
	1/6th LFs move to No. 1 Australian Line.

—

9–13 August	1/6th LFs in No. 1 Australian Line.
13–19 August	42nd Division (including 1/6th LFs) in corps reserve at bivouac area near X Beach.
19 August	42nd Division relieves the 29th Division in the 'Left Section.' *(See map 8.)*
	1/6th LFs relieve 4th Worcester Regt on extreme right of the division's firing-line, from a point left of Border Barricade to the divisional boundary with the RND on the right.
	Lord Rochdale arrives back from England.
22 August	1/6th LFs in dug-outs at Geoghegan's Bluff in brigade reserve.
25 August	1/6th LFs at Gully Beach in corps reserve.
2 September	1/6th LFs in firing and support lines above Fusilier Bluff.
5 September	1/6th LFs in brigade reserve: A & B coys in Trolley Ravine, C & D coys and Bn HQ in Y Ravine.
8 September	1/6th LFs in firing and support lines above Fusilier Bluff.

11 September	1/6th LFs in brigade reserve: C & D coys in Trolley Ravine, A & B coys and Bn HQ in Y Ravine.
18 September	1/6th LFs at Gully Beach in corps reserve.
24 September	1/6th LFs in firing and support lines from Border Barricade (inclusive) to Union Street, on right of Gully Ravine.
29 September	Lord Rochdale sick to hospital, then to hospital ship and evacuated.
1 October	42nd Division reorganised into two brigades of six under-strength battalions: Group 1, made up of 127th Bde, 1/4th East Lancs Regt, and 1/8th LFs; and Group 2, made up of three battalions each of the 125th and 126th Bdes.

1/6th LFs (as part of Group 2) at Gully Beach in corps reserve. |
| 15 October | Group 2 takes over the Right Sub-section. (The British sector was divided into three sections: Left Section, Centre Section and Right Section. Each section was divided again into a Left Sub-section and Right Sub-section.)

1/6th LFs in firing and support lines from Border Barricade (inclusive) to Northern Barricade (divisional boundary) on the right. |
| 19 October | 1/6th LFs in Douglas Street and Fusilier Street as sub-section reserve. |
| 21 October | Due shortage of officers, 1/6th LFs split up: two coys to 1/5th LFs and two coys to 1/7th LFs.

1/6th LFs relieves 1/5th & 1/7th LFs, taking over entire Right Sub-section firing-line. |

29 October 1/7th LFs, with two coys 1/6th LFs, at Geoghegan's Bluff and Douglas and Fusilier Streets.

1/5th LFs, with two coys 1/6th LFs, at Gully Beach.

2 November 1/6th LFs men in Douglas and Fusilier Streets move to Geoghegan's Bluff.

One officer and 65 other ranks proceed to Mudros for training.

12 November 125th Bde reorganised as two battalions: 1/5th & 1/8th LFs with 50 men of 1/6th LFs, as one battalion under Lt. Col. F. A. Woodcock; and remainder of 1/6th LFs with 1/7th LFs as one battalion under Lt. Col. C. T. Alexander.

125th Bde takes over Right Sub-section.

1/6th LFs dispersed between various battalions in firing-line, support line and in Douglas and Fusilier Streets.

20 November Divisional boundary moves to a point 20 yards east of Union Street (firing-line) and a point 20 yards east of Diggle Street (support line).

26 November 125th Bde relieved and goes into divisional reserve.

1/5th LFs plus two officers and 90 other ranks from 1/6th LFs in Y Ravine.

Remainder of 1/6th LFs in Eski Line East (the section of the Eski Line east of Gully Ravine).

10 December 125th Bde takes over the Right Sub-section.

1/6th LFs dispersed between various battalions in firing-line support line and in Frith Walk and Douglas and Fusilier Streets.

Diversionary Attack against the Gridiron

The Gridiron was an enemy position on the east side of Gully Ravine opposite the Birdcage Walk firing-line. See map 8.

19 December	Three diversionary attacks were carried out at Helles on 19 December with the intention of distracting enemy attention from Anzac and Suvla, on the eve of the evacuation of both zones.
	The 1/7th LFs carried out a successful attack on the Gridiron supported by 1/6th LFs (less 30 men then with 1/5th & 1/8th LFs).
14:15	1/7th LFs attack goes forward with 1/6th LFs holding the firing-line. 1/6th LFs also provided two grenade parties (10 men) at Brennan's Post, and manned the section's grenade catapults. The remainder of 1/6th LFs in support and reserve trenches.
19–22 December	Fighting continued but 1/7th LFs and 1/6th LFs (with assistance from South Eastern Mounted Brigade grenade teams) repulsed all enemy counter-attacks and held onto the ground gained.

—

24 December	125th Brigade relieved from the Right Sub-section and goes into corps reserve.
	1/6th LFs in winter quarters at 'Sites 6 & 7, Gully Ravine.'
	Battalion machine-guns in Eski Line West.
27 December	125th Brigade embarks from W Beach (Lancashire Landing) on SS *Ermine* for Mudros.

—

Officers of the 1/6th Lancs Fusiliers.

(This portrait was taken in Cairo, late 1914 or early 1915.)

6TH BATTALION

First Row — Left to Right — Capts. E. Woolmer, G. Scott, A. V. Clegg, Major Heywood, Major Lees, Lt.-Col. Lord Rochdale, Capt. and Adjutant Spafford, Capt. Gledhill, Capt. Barker, Capt. Crossley.

Second Row — Lt. L. M. Robinson, 2nd Lt. T. D. M. Bartley, Lt. G. G. Holden, Lt. Brentnall, 2nd Lt. Roe, 2nd Lt. Jones, 2nd Lt. Hornby, Lt. J. S. Lord, Lt. N. V. Holden, Lt. O'Neill, 2nd Lt. Mathews.

Third Row — 2nd Lt. E. Duckworth, Lt. R. W. Leach, Capt. and Q.-Mr. Griffiths, 2nd Lt. P. V. Davies, 2nd Lt. G. Wyatt, 2nd Lt. C. B. Storey, Lt. J. H. Smith, 2nd Lt. G. E. Marriatt, 2nd Lt. M. Carton de Wiart.

Promotions — Major Lees, now Lt.-Col.; Capt. Gledhill, now Major; Lt. Roe, now Capt.; 2nd Lt. Hornby, now Capt.; Lt. J. S. Lord, now Capt.

APPENDIX VII

Establishment, Drafts & Battle Casualties, 1/6th Lancashire Fusiliers at Gallipoli

Establishment on landing at V Beach on 5 May 1915

32 officers, 878 other ranks and two machine-guns.

Drafts of reinforcements

Taken from the 125th Brigade HQ and 42nd Division 'A' Branch diaries. Some small drafts from England or Egypt may have gone unrecorded. The list does not include men rejoining from hospital.

31 May	Capt. P. C. Joyce (Connaught Rangers — attached Egyptian Army)
21 June	One officer (2nd Lieut. T. H. Coe) and 95 other ranks from England
23 July	Four officers and 50 other ranks from England
20 August	Two officers and 92 other ranks from England
7 October	Five officers from England
22 October	Five officers and 148 other ranks arrive from England
27 October	Eleven other ranks (possibly from Egypt or hospital)
2 November	One officer and 75 other ranks arrive from England
23 December	Lieut. Brentnall, RAMC, arrives from England

Total for officers: 31
Total for other ranks: 471

Battle Casualties

Second Battle of Krithia

Taken from the 125th Brigade HQ and 42nd Division 'A' Branch diaries.

Officers killed:	Nil
Other ranks killed:	14
Officers wounded:	Capt. G. Scott
	Capt. J. J. Gledhill
	Lieut. M. Robinson
	2nd Lieut. P. V. Davis (*sic* Davies)
	2nd Lieut. M. C. De Wiart
	2nd Lieut. J. S. D. Berrington
Other ranks wounded:	152
Other ranks missing:	43

Third Battle of Krithia

Casualties between 4–9 June, taken from the 125th Brigade HQ and 42nd Division 'A' Branch diaries.

Officers killed:	Lieut. N. V. Holden
Other ranks killed:	26
Officers wounded:	Lieut. T. R. Taylor
	Lieut. R. W. Leach
Other ranks wounded:	193
Other ranks missing:	2

Battle of the Vineyard

Casualties between 6–8 August taken from the 42nd Division 'A' Branch diary and CWGC records.

Officers killed:	Capt., & Adjutant A. L. Spafford
	Capt., & Quartermaster W. H. Griffiths, DCM
	2nd Lieut. E. Duckworth
Other ranks killed:	58
Officers wounded:	Capt. A. V. Clegg
	Capt. L. R. Rowe (9th Lincoln Regt)
	Lieut. O. Cooper
	2nd Lieut. R. Leak
	2nd Lieut. W. Redmond
Other ranks wounded:	Unknown
Other ranks missing:	Unknown

Action at the Gridiron on 19 December 1915

Casualties between 19–22 December, taken from the 42nd Division 'A' Branch diary and CWGC records.

Officers killed:	Nil
Other ranks killed:	13
Officers wounded:	One unknown officer
Other ranks wounded:	27
Other ranks missing:	Unknown

APPENDIX VIII

Gallipoli Roll of Honour

1/6th Lancashire Fusiliers

Key

Double dagger symbol ‡ indicates soldiers with records showing both 1/6th and 1/8th Lancs Fusiliers.

Information [in square brackets] assumed and not fully substantiated.

MIC DoD: Date of death according to Medical Index Card.

CWGC DoD: Date of death according to the records of the Commonwealth War Graves Commission.

Cemeteries and memorials (at Gallipoli unless otherwise indicated):

AC	Alexandria (Chatby) Cemetery, Egypt
CWMC	Cairo War Memorial Cemetery, Egypt
EMMC	East Mudros Military Cemetery, Lemnos
HM	Helles Memorial
LLC	Lancashire Landing Cemetery
NMC	Netley Military Cemetery, England
PF	Pink Farm Cemetery
PMC	Pieta Military Cemetery, Malta
PMCL	Portianos Military Cemetery, Lemnos
RC	Redoubt Cemetery
SBC	Skew Bridge Cemetery
TTC	Twelve Tree Copse Cemetery

Manner of death, shown in brackets after the cemetery reference:

K	Killed in action
D	Died of wounds
d	Died of sickness

References

Medal Index Cards, WW1 Medal and Award Rolls, Pension records and *Soldiers Died in the Great War, 1914–1919*.

Gallipoli Roll of Honour, 1/6th Lancashire Fusiliers, Other Ranks, in alphabetical order

Name, number, rank & age	Date landed	Birthplace	Place of enlistment	MIC DoD	CWGC DoD	Cem/Mem
Ashurst, Thomas, 9420, Pte (23)	5.05.15	Rochdale	Rochdale	7.08.15	7.08.15	HM (K)
Ashworth, Fred, 9548, Pte (19)	5.05.15	Castleton	Rochdale	4.06.15	4.06.15	HM (K)
Ashworth, John, 9628, Pte	5.05.15	Bury	Rochdale	11.07.15	11.07.15	AC (D)
Atkinson, Clifford, 9056, Pte (17)	5.05.15	Bradford	Middleton	13.05.15	13.05.15	HM (K)
Bailey, William, 6696, Cpl (30)	5.05.15	Todmorden	Todmorden	6.05.15	6.05.15	HM (K)
Ball, Frederick, 9240, Cpl (33)	5.05.15	Preston	Rochdale	15.05.15	8.05.15	HM (K)
Barker, Frank Astin, 9700, Pte (17)	5.05.15	Todmorden	Todmorden	8.05.15	8.05.15	TTC (K)
Beaman, Percy, 9498, Pte (21)	5.05.15	Middleton	Middleton	8.05.15	8.05.15	HM (K)
Birbeck, James, 9548, Pte (19)	5.05.15	Rochdale	Rochdale	4.06.15	4.06.15	HM (K)
Black, Lauchlan, 8567, Pte	20.08.15	Wardle	Rochdale	21.12.15	21.12.15	LLC (D)
Boothman, James Walter, 6648, Sgt (30)	5.05.15	Rochdale	Rochdale	18.07.15	18.07.15	LLC (K)
Bridge, Frank, 9512, Pte (18)	5.05.15	[Rochdale]	Rochdale	31.07.15	31.07.15	LLC (K)
Brierley, John William, 8584, Pte (20)	5.05.15	Rochdale	Rochdale	4.06.15	6.06.15	TTC (K)

ROLL OF HONOUR

Name, number, rank & age	Date landed	Birthplace	Place of enlistment	MIC DoD	CWGC DoD	Cem/Mem
Brierley, Norman Clifford, 8909, Pte	5.05.15	Rochdale	Littleborough	10.11.15	10.11.15	HM (K)
Broadbent, Samuel, 7019, L/Sgt (Cpl)	5.05.15	Rochdale	Rochdale	17.08.15	17.08.15	AC (D)
Brookes, James, 9496, Pte (27)	5.05.15	Irlam	Middleton	7.08.15	7.08.15	HM (K)
Brotherton, Samuel, 9596, Pte (25)	5.05.15	Rochdale	Rochdale	7.08.15	7.08.15	HM (K)
Broxton, Edwin, 8797, Pte (19)	5.05.15	Manchester	Rochdale	4.07.15	5.06.15	HM (D)
Burgess, James, 7079, Sgt (31)	5.05.15	Stockport	Middleton	8.05.15	8.05.15	HM (K)
Burrows, James, 7515, Pte (29)	5.05.15	Rochdale	Rochdale	26.07.15	26.07.15	AC (d)
Butterworth, Wilson, 10088, Pte (25)	22.07.15	Errington	Todmorden	7.08.15	7.08.15	HM (K)
Callon, John William, 9103, Pte (30)	5.05.15	Manchester	Todmorden	23.06.15	1.06.15	AC (D)
Carpenter, William Henry, 9733, Pte	5.05.15	[Todmorden]	Todmorden	15.07.15	15.07.15	RC (K)
Chadwick, Charles Henry, 9650, Pte (34)	5.05.15	Manchester	Rochdale	1.08.15	1.08.15	HM (K)
Chadwick, Herbert, 9163, Pte (27)	5.05.15	Rochdale	Rochdale	7.08.15	7.08.15	HM (K)
Child, John Willie, 8978, Pte DCM (17?)	5.05.15	Failsworth	Todmorden	4.08.15	4.08.15	LLC (D)
Clarke, Edward, 9632, Pte (18)	5.05.15	Todmorden	Rochdale	4.07.15	6.06.15	HM (K)
Clarke, Herbert, 8043, Sgt (23)	5.05.15	Halifax	Todmorden	18.06.15	5.06.15	LLC (D)
Clegg, Albert, 8492, Pte (21)	5.05.15	Wardle	Rochdale	7.06.15	12.05.15	HM (K)
Connolly, Frank, 9502, Pte (19)	5.05.15	[Middleton]	Middleton	4.07.15	4.06.15	HM (K)

APPENDIX VIII

Name, number, rank & age	Date landed	Birthplace	Place of enlistment	MIC DoD	CWGC DoD	Cem/Mem
Connolly, Thomas, 9397, Pte (26)	5.05.15	[Middleton]	Middleton	4.07.15	4.06.15	HM (K)
Consterdine, Fred, 9044, Pte (18)	5.05.15	Blackley	Middleton	7.08.15	7.08.15	HM (K)
Cook, Robert, 6991, L/Sgt/Cpl (26)	5.05.15	Shaw	Rochdale	5.06.15	5.06.15	HM (K)
Cook, Sidney James, 9774, Pte	5.05.15	Burnley	Todmorden	29.05.15	29.05.15	HM (D)
Crowther, Harold, 9678, Pte (26)	5.05.15	Todmorden	Todmorden	15.05.15	15.05.15	HM (K)
Davies, Chadwick, 8278, Cpl (23)	5.05.15	Middleton	Middleton	20.12.15	20.12.15	TTC (K)
Dawson, David, 8526, L/Cpl (21)	5.05.15	Swansea	Todmorden	2.10.15	2.10.15	PMCL (d)
Dean, Arthur, 8290, Pte (23)	22.07.15	Middleton	Middleton	7.08.15	7.08.15	RC (K)
Dean, James, 6847, Sgt (30)	5.05.15	Middleton	Middleton	7.08.15	7.08.15	RC (K)
Dearden, Ernest, 8398, Pte (23)	5.05.15	Rochdale	Rochdale	6.06.15	6.06.15	HM (K)
Donegani, Joseph, 10204, Pte (21)	21.06.15	Rochdale	Rochdale	7.08.15	7.08.15	HM (K)
Duckworth, Samuel, 9983, Pte (33)	22.07.15	Rochdale	Rochdale	7.09.15	7.09.15	HM (K)
Dunn, Frank, 10459, Pte (24)	22.07.15	Rochdale	Rochdale	7.08.15	7.08.15	HM (K)
Edmonson, Willie, 9738, Pte (20)	5.05.15	Hadfield	Todmorden	17.06.15	13.06.15	HM (K)
Ellis, George, 8394, Pte (23)	5.05.15	Hebden Bdge	Todmorden	15.05.15	7.05.15	HM (K)
Ellison, John Willie, 7814, Pte (23?)	5.05.15	Walsden	Todmorden	10.06.15	10.06.15	LLC (D)
English, Patrick, 9997, Pte (32)	22.07.15	Rochdale	Rochdale	14.08.15	14.08.15	HM (D)

Name, number, rank & age	Date landed	Birthplace	Place of enlistment	MIC DoD	CWGC DoD	Cem/Mem
Farmer, Ernest, 10357, Pte (35)	21.06.15	Rochdale	Rochdale	7.08.15	7.08.15	HM (K)
Farrar, Walter, 9679, Pte (20)	5.05.15	Todmorden	Todmorden	20.05.15	14.06.15	PMC (D)
Fielder, Fred, 10067, Pte	22.07.15	Todmorden	Todmorden	7.08.15	7.08.15	HM (K)
‡ Fisher, Edward (Edwin) 8672, Pte	5.05.15	Rochdale	Rochdale	14.05.15	14.05.15	LLC (D)
Fitton, Arnold, 9996, Pte (18)	21.06.15	Rochdale	Rochdale	7.09.15	7.08.15	HM (K)
Foxall, Arthur, 9463, Pte (24)	5.05.15	Rhodes, Lanc	Middleton	10.11.15	10.11.15	HM (K)
Gibson, Thomas, 8008, Sgt (28)	5.05.15	Todmorden	Rochdale	7.08.15	7.08.15	HM (K)
Goodier, Alfred, 9379, Pte (23)	5.05.15	Littleborough	Rochdale	16.05.15	16.05.15	AC (D)
Gordon, Fred, 8717, Pte (20)	5.05.15	Middleton	Middleton	4.06.15	4.06.15	HM (K)
Gosling, John, 8940, Pte (18)	5.05.15	Macclesfield	Middleton	30.09.15	30.09.15	HM (K)
Graham, Robert, 9201, Pte (41)	5.05.15	Harrogate	Middleton	4.07.15	4.06.15	HM (K)
Grannan, Thomas, 8687, Pte (20)	5.05.15	Todmorden	Todmorden	27.12.15	27.12.15	LLC (D)
Greenwood, Ernest, 8284, Pte (21)	5.05.15	Todmorden	Todmorden	20.12.15	20.12.15	HM (K)
‡ Greenwood, Ernest, 8841, Pte	5.05.15	Todmorden	Todmorden	4.09.15	7.08.15	HM (K)
Greenwood, Harold, 9695, Pte (17)	5.05.15	Todmorden	Todmorden	21.06.15	21.06.15	HM (K)
‡ Greenwood, Thomas, 9652, Pte	5.05.15	Todmorden	Todmorden	4.09.15	7.08.15	HM (K)
Greenwood, Wm. Albert, 9720, Pte (19)	5.05.15	Todmorden	Todmorden	20.08.15	20.08.15	LLC (K)

APPENDIX VIII

Name, number, rank & age	Date landed	Birthplace	Place of enlistment	MIC DoD	CWGC DoD	Cem/Mem
Grimshaw, Jocks, 9460, Pte (19)	5.05.15	[Middleton]	Middleton	19.06.15	19.06.15	HM (K)
Guthrie, Thomas, 6925, Sgt (38)	5.05.15	Belfast	Rochdale	7.08.15	7.08.15	HM (K)
Halsall, George Thomas, 8961, Pte	5.05.15	Castleton	Rochdale	15.05.15	15.05.15	HM (K)
Hamer, Herbert, 9402, Pte (25)	5.05.15	[Middleton]	Middleton	7.08.15	7.08.15	HM (K)
Haney, John, 9557, Pte	5.05.15	Middleborough	Rochdale	8.08.15	8.08.15	RC (K)
Hardiker, Harry, 9334, L/Cpl (Pte)	5.05.15	[Rochdale]	Rochdale	9.08.15	9.08.15	HM (K)
Hazeltine, Herbert, 8346, L/Cpl (Pte)	5.05.15	Todmorden	Todmorden	5.06.15	5.06.15	HM (K)
Henderson, Edward, 8922, Pte (19)	5.05.15	Prestwich	Rochdale	4.07.15	4.06.15	RC (K)
Henfrey, Harold, 9106, Pte (20/21)	5.05.15	[Rochdale]	Rochdale	18.06.15	5.06.15	LLC (D)
Heys, Herbert, 8618, Pte (22)	5.05.15	Rawtenstall	Rochdale	3/4.07.15	21.06.15	HM (K)
‡ Heywood, Elijah, 9482. Pte (21)	5.05.15	Middleton	Middleton	7.08.15	7.08.15	HM (K)
Hollows, Brinton, 9073, Pte (17)	5.05.15	Halifax	Rochdale	26.07.15	26.07.15	LLC (D)
Hopkinson, John, 8830, Pte (21)	5.05.15	Manchester	Rochdale	4.07.15	4.06.15	RC (K)
Howarth, Samuel, 9023, Pte	21.06.15	Whitworth	Rochdale	12.08.15	12.08.15	HM (K)
Hoyle, Joseph, 9509, Pte	5.05.15	Rochdale	Rochdale	8.08.15	8.08.15	EMMC (D)
Humphreys, Frederick, 9501, Pte	5.05.15	Chorley	Middleton	8.08.15	9.08.15	LLC (D)
Hunt, John, 9920, Pte (26)	21.06.15	Middleton	Middleton	7.08.15	7.08.15	HM (K)

Name, number, rank & age	Date landed	Birthplace	Place of enlistment	MIC DoD	CWGC DoD	Cem/Mem
Hyland, Edward, 9453, Pte	5.05.15	Heywood	Rochdale	7.08.15	7.08.15	RC (K)
Ingham, Samuel, 9744, L/Cpl (Pte)	5.05.15	Hindley	Rochdale	20.12.15	20.12.15	TTC (K)
Izat, James, 8669, Sgt (20)	5.05.15	Rochdale	Rochdale	23.07.15	23.07.15	HM (D)
Jackson, Evelyn, 10786, Pte (18)	22.10.15	Rochdale	Rochdale	15.12.15	15.12.15	AC (d)
Jacques, Robert, 8812, Pte (22)	5.05.15	Middleton	Middleton	7.08.15	7.08.15	HM (K)
Jagger, Harold, 8588, Pte (21)	5.05.15	Huddersfield	Rochdale	4.07.15	6.06.15	HM (K)
Jeffrey, Robert, 9597, Pte (28)	5.05.15	Rochdale	Rochdale	4.07.15	4.06.15	RC (K)
Johnson, Albert Edward, 9218, Sgt (44)	5.05.15	Lincoln	Rochdale	7.08.15	7.08.15	HM (K)
Jones, William, 9538, Pte (37)	5.05.15	Littleborough	Rochdale	7.08.15	7.08.15	HM (K)
Kelsey, John, 9152, A.L/Cpl (20/21)	5.05.15	Manchester	Todmorden	30.06.15	10.05.15	HM (D)
Kenyon, Wm. Henry, 9510, Pte (23)	5.05.15	Shaw	Rochdale	15.05.15	15.05.15	HM (K)
King, James, 8835, Pte (20)	5.05.15	Hollinwood	Todmorden	4.07.15	5.06.15	RC (K)
Knott, James, 10728, Pte	22.10.15	Athlone	Rochdale	18.11.15	18.11.15	PF (D)
Langley, William, 10242, Pte	21.06.15	Normanton	Rochdale	19.12.15	19.12.15	TTC (K)
Law, James Wm. 8512, Pte (20/21)	5.05.15	Todmorden	Rochdale	10.06.15	10.06.15	LLC (D)
Lawrence, James, 9140, Pte (19)	5.05.15	Newcastle	Todmorden	4.06.15	4.06.15	RC (K)
Lawton, John, 9536, Pte (28)	5.05.15	[Rochdale]	Rochdale	18.06.15	7.06.15	HM (D)

APPENDIX VIII

Name, number, rank & age	Date landed	Birthplace	Place of enlistment	MIC DoD	CWGC DoD	Cem/Mem
Lees, Harold, 9487, Pte (23)	5.05.15	Stalybridge	Middleton Junc.	3.09.15	3.09.15	TTC (K)
Lewis, Ernest, 8656, Pte (19)	5.05.15	Walsden	Todmorden	22.06.15	22.06.15	LLC (K)
Lewis, Fred, 8958, Pte (17)	5.05.15	Middleton	Middleton	7.08.15	7.08.15	RC (K)
Liddle, Alfred, 8762, Cpl (20)	5.05.15	Middleton	Middleton	7.08.15	7.08.15	HM (K)
Livesey, Richard, 8504, Pte (21)	5.05.15	Todmorden	Todmorden	7.08.15	7.08.15	HM (K)
Livsey, James, 9026, Pte (21)	21.06.15	Heywood	Rochdale	7.08.15	7.08.15	HM (K)
Lomax, James, 9371, Pte (21)	5.05.15	[Rochdale]	Rochdale	9.06.15	4.06.15	HM (K)
MacDonald, George Angus, 8559, Pte	21.06.15	Middleton	Rochdale	7.08.15	7.08.15	HM (K)
Marland, William, 9605, Pte (20)	5.05.15	Rochdale	Rochdale	12.08.15	12.08.15	HM (D)
Marsden, Joseph, 9273, L/Sgt (39)	5.05.15	Oldham	Rochdale	10.09.15	10.09.15	AC (d)
Mason, John, 4248, CSM. (47)	5.05.15	Manchester	Todmorden	7.08.15	7.08.15	LLC (K)
McGibney, John, J. 8620, Pte (20)	5.05.15	Castleton	Rochdale	15.05.15	15.05.15	HM (K)
Mellor, John, 7646, Sgt (26)	5.05.15	Bury	Rochdale	21.12.15	21.12.15	HM (K)
Mills, Harold, 9874, Pte (20)	21.06.15	Rochdale	Rochdale	7.08.15	7.08.15	HM (K)
Milne, Harry, 9580, Pte (20)	5.05.15	Castleton	Rochdale	19.12.15	8.08.15	HM (K)
Mitchell, Jas. Herbert, 8847, L/Cpl (21)	5.05.15	Todmorden	Todmorden	12.08.15	12.08.15	LLC (K)
Moran, Thomas, 10373, Pte	21.06.15	[Rochdale]	Rochdale	7.08.15	7.08.15	HM (K)

Name, number, rank & age	Date landed	Birthplace	Place of enlistment	MIC DoD	CWGC DoD	Cem/Mem
Morris, Alfred, 8424, Pte	5.05.15	Shaw	Rochdale	8.08.15	8.08.15	HM (K)
Morrow, Alexander, 7338, Sgt (28)	5.05.15	Belfast	Middleton	8.08.15	8.08.15	RC (K)
Mudd, James, 8119, Pte (28)	5.05.15	Oldham	Rochdale	4.06.15	4.06.15	RC (K)
Murphy, Frank, 10282, Pte (17)	5.05.15	Rawtenstall	Rochdale	12.07.15	12.07.15	HM (D)
Myatt, John, 8645, Pte (19)	5.05.15	Knutsford	Rochdale	4.07.15	4.06.15	HM (K)
Naylor, John, 9436, Pte (20)	5.05.15	[Rochdale]	Rochdale	15.06.15	15.06.15	EMMC (D)
Nelson, Wm. Thomas, 7273, Sgt (35)	5.05.15	Shaw	Rochdale	15.05.15	15.05.15	TTC (D)
Newton, John, 10889, Pte (19)	22.10.15	Manchester	Rochdale	20.12.15	20.12.15	TTC (K)
Nicholl, Frank, 10289, Pte (21)	21.06.15	Walsden	Todmorden	7.08.15	7.08.15	HM (K)
Ogden, Jas Robert Stott, 7630, Sgt (29)	5.05.15	Rochdale	Rochdale	14.07.15	14.07.15	RC (K)
Oliver, Leonard, 8263, Pte (27)	21.06.15	Middleton	Rochdale	7.08.15	7.08.15	HM (K)
Osbaldeston, Geo. Kenyon, 9567, Pte	5.05.15	[Rochdale]	Rochdale	7.08.15	7.08.15	HM (K)
Parry, Wilfred, 9617, Pte (25)	5.05.15	Rochdale	Rochdale	11.06.15	11.06.15	PMC (D)
Partington, James, 9278, Pte	22.07.15	Middleton	Rochdale	7.08.15	7.08.15	RC (K)
Pearson, James, 9558, Pte (23)	5.05.15	Bury	Rochdale	4.07.15	6.06.15	HM (K)
Phillips, John, 9712, Pte	5.05.15	Wellington	Todmorden	7.06.15	8.05.15	HM (K)
Pickering, Harold, 10703, Pte (21)	5.05.15	Worsley	Egypt	15.05.15	15.05.15	HM (K)

APPENDIX VIII

Name, number, rank & age	Date landed	Birthplace	Place of enlistment	MIC DoD	CWGC DoD	Cem/Mem
Pickles, Ernest, 10238, Pte (38)	21.06.15	Todmorden	Rochdale	7.08.15	7.08.15	HM (K)
Pickles, Frederick, 10615, Pte	22.07.15	Todmorden	Rochdale	7.08.15	7.08.15	HM (K)
Pollard, Walter, 10261, Pte	21.06.15	Rochdale	Rochdale	7.08.15	7.08.15	HM (K)
Pollitt, John, 9448, Pte (22)	5.05.15	[Rochdale]	Rochdale	22.06.15	22.06.15	HM (D)
Porter, William, 8831, Pte (20)	5.05.15	Rochdale	Rochdale	7.08.15	7.08.15	HM (K)
Powell, George, 9703, Pte	5.05.15	Harrogate	Todmorden	7.08.15	7.08.15	HM (K)
Price, Samuel James, 8517, Pte	5.05.15	Lancaster	Rochdale	6.05.15	6.05.15	HM (K)
Radcliffe, Frank, 10897, Pte (37?)	1.10.15	Bamford	Rochdale	25.11.15	25.11.15	PMC (d)
Ratcliffe, Thomas, 8757, Pte (23?)	5.05.15	Middleton	Todmorden	12.08.15	12.08.15	RC (K)
Richardson, William, 10196, Pte	5.05.15	Failsworth	Middleton	7.08.15	7.08.15	HM (K)
Rigby, Thomas, 7964, Cpl (26)	5.05.15	Southport	Rochdale	23.06.15	16.05.15	HM (d)
Riley, Frank, 9128, Pte (20)	5.05.15	Rochdale	Rochdale	7.08.15	7.08.15	HM (K)
Risby, Richard, 10827, Pte (18)	22.10.15	Rochdale	Rochdale	28.12.15	28.12.15	AC (D)
Roberts, Wm. Henry, 9282, Pte (34)	5.05.15	Oldham	Rochdale	7.08.15	7.08.15	HM (K)
Rogers, Denis, 9640, Cpl (37)	5.05.15	[Burnley]	Todmorden	7.08.15	7.08.15	HM (K)
Ryle, William, 10497, Pte	21.06.15	[Rochdale]	Rochdale	22.12.15	22.12.15	TTC (D)

Name, number, rank & age	Date landed	Birthplace	Place of enlistment	MIC DoD	CWGC DoD	Cem/Mem
Scholes, William, 8315, L/Cpl (22)	5.05.15	Lovell, Mass. USA	Rochdale	22.12.15	22.12.15	TTC (K)
Scott, Elijah, 8115, Cpl (24)	5.05.15	Rochdale	Rochdale	15.08.15	15.08.15	HM (D)
Senior, Harry, 8303, Pte (22)	5.05.15	Bolton	Middleton	7.08.15	7.08.15	HM (K)
Sharpe, Maurice, 9569, Pte (24)	5.05.15	[Rochdale]	Rochdale	7.08.15	7.08.15	HM (K)
Sharpe, Wm. Alfred, 8826, Pte (18?)	5.05.15	Castleton	Rochdale	4.06.15	5.06.15	HM (K)
Sheard, James, 10160, Pte (36)	21.06.15	Castleton	Rochdale	8.08.15	8.08.15	HM (K)
Shepherd, James, 10270, Pte (24)	22.07.15	Rochdale	Rochdale	7.08.15	7.08.15	HM (K)
Shotton, Chas. Henry, 9618, Pte (24)	5.05.15	Rochdale	Rochdale	5.07.15	21.06.15	HM (K)
Smith, Arthur Wm., 9382, L/Cpl (24)	5.05.15	[Castleton]	Rochdale	7.08.15	7.08.15	HM (K)
Smith, Leonard, 7924, L/Cpl (26?)	5.05.15	Leominster	Rochdale	4.06.15	4.06.15	HM (K)
Smith, William, 9750, Pte	5.05.15	[Rochdale]	Rochdale	12.05.15	12.05.15	HM (D)
Smith, William, 10000, Pte (22)	21.06.15	Todmorden	Todmorden	7.08.15	7.08.15	HM (K)
Smith, Willie, 8408, Pte	5.05.15	Walsden	Todmorden	19.06.15	4.06.15	HM (K)
Southwell, Harold, Barston, 9692, Pte	5.05.15	[Todmorden]	Todmorden	18.08.15	14.08.15	HM (D)
Spencer, Thomas, 7947, Cpl (26)	5.05.15	Wardleworth	Rochdale	16.07.15	16.07.15	LLC (D)
Stock, Albert, 9524, Pte (38)	5.05.15	[Rochdale]	Rochdale	7.06.15	8.05.15	HM (K)

APPENDIX VIII

Name, number, rank & age	Date landed	Birthplace	Place of enlistment	MIC DoD	CWGC DoD	Cem/Mem
Stockton, John Wm., 9570, Pte (29)	5.05.15	[Rochdale]	Rochdale	7.08.15	7.08.15	HM (K)
Styles, Geo. Edward, 10839 Pte (31)	1.10.15	Rochdale	Rochdale	21.12.15	21.12.15	TTC (K)
Sutcliffe, William, 9722, Pte	5.05.15	Todmorden	Todmorden	5.06.15	5.06.15	HM (K)
Sutcliffe, William, 10980, Pte	22.10.15	Littleborough	Rochdale	14.11.15	14.11.15	LLC (D)
Swaine, Arthur, 9335, Pte (33)	5.05.15	[Rochdale]	Rochdale	15.05.15	8.05.15	HM (K)
Sykes, Chas. Edward, 9337, Pte (18?)	5.05.15	Rochdale	Rochdale	4.07.15	6.06.15	RC (K)
Taylor, Henry Hardman, 8565, Pte (21)	5.05.15	Rochdale	Rochdale	19.06.15	19.06.15	LLC (D)
Taylor, Richard Royds, 9074, Pte (20)	5.05.15	Rochdale	Rochdale	6.05.15	6.05.15	HM (K)
Taylor, Walter, 9986, Pte	5.05.15	Rochdale	Rochdale	7.08.15	7.08.15	EMMC (d)
Thomas, Charles Wm, 8886, Pte (22)	5.05.15	Todmorden	Todmorden	15.05.15	15.05.15	TTC (K)
Tillotson, James, 10245, Pte	21.06.15	Colne	Rochdale	7.08.15	7.08.15	HM (K)
Turles, Herbert, 9493, Pte	5.05.15	Middleton	Middleton	5.06.15	5.06.15	HM (K)
Turner, Chas. Edward, 10106, Pte (22)	22.07.15	Rochdale	Rochdale	7.08.15	7.08.15	HM (K)
Turner, Frank, 9581, L/Cpl (20)	5.05.15	Rochdale	Rochdale	19.12.15	19.12.15	TTC (K)
Wacey, Willie, 9611, Pte (26)	5.05.15	E. Dereham	Rochdale	7.08.15	7.08.15	HM (K)
Walton, William, 10027, Pte (19)	21.06.15	Todmorden	Todmorden	10.08.15	8.08.15	LLC (D)

ROLL OF HONOUR

Name, number, rank & age	Date landed	Birthplace	Place of enlistment	MIC DoD	CWGC DoD	Cem/Mem
Ward, James, 9593, Pte (52?)	5.05.15	Wolverhampton	Rochdale	4.07.15	4.06.15	HM (K)
Webster, Henry, 7554, Sgt (23?)	5.05.15	Rochdale	Rochdale	15.05.15	8.05.15	HM (K)
Wellens, William, 8787, L/Cpl (27)	5.05.15	Middleton	Middleton	8.08.15	8.08.15	RC (K)
Whitham, Marshall, 8667, Pte (22)	5.05.15	Todmorden	Todmorden	3.06.15	3.06.15	SBC (K)
Whittaker, William, 10454, Pte (42)	22.07.15	[Rochdale]	Rochdale	7.08.15	7.08.15	HM (K)
Whittaker, William, 10614, Pte (25)	22.07.15	Rochdale	Rochdale	20.12.15	20.12.15	TTC (K)
Whitworth, John, 9403, Pte (43)	5.05.15	Cardiff	Middleton	14.05.15	4.07.15	NMC (D)
Wild, George, 8668, L/Cpl (21)	5.05.15	Royton	Rochdale	13.01.16	20.12.15	LLC (D)
Wild, George, 7772, Cpl (25)	5.05.15	Middleton	Middleton	7.08.15	7.08.15	HM (K)
Wild, Joe, 9032, Pte	5.05.15	Royton	Rochdale	25.06.15	20.06.15	LLC (D)
Wildblood, Ralph, 8425, Pte	5.05.15	Rochdale	Rochdale	4.07.15	4.06.15	HM (K)
Williams, Francis, Littleton, 9110, Pte (24)	5.05.15	Lostwithiel	Todmorden	19.06.15	4.06.15	HM (K)
Wilson, James, 9354, Pte (25)	5.05.15	[Rochdale]	Rochdale	4.07.15	4.06.15	HM (K)

Total other ranks: 193

Gallipoli Roll of Honour, 1/6th Lancashire Fusiliers, Officers

Rank, name & age	Date landed	Birthplace	Place of enlistment	MIC DoD	CWGC DoD	Cem/Mem
Capt. A. Victor Clegg (30)	5.05.15	Littleborough	Rochdale	7.08.15	7.08.15	LLC (D)
Capt. & QM Wm. Henry Griffiths (55)	5.05.15	Horton, Middx	London?	Aug 15	7.08.15	HM (K)
Capt. Arthur Langworthy Spafford (Adj, 36) – 1st LF, attached	5.05.15	Bowdon	Bury?	7.08.15	7.08.15	LLC (D)
Lieut. Frederick Wm. Harvey (35) – 10th South Lancs Regt, attached	5.05.15?	Hull	Crosby?	9.08.15	10.08.15	LLC (D)
Lieut. Norman Victor Holden (25)	5.05.15	Harpurhey	Rochdale	4.06.15	4.06.15	LLC (D)
Lieut. Samuel O'Neill (21)	5.05.15	Castleton	Rochdale	12.06.15	12.06.15	LLC (D)
Lieut. Joshua Harold Smith (24)	5.05.15	Southport	Rochdale	12.08.15	12.08.15	RC (K)
2nd Lieut. Eric Duckworth (19)	5.05.15	Rochdale	Rochdale	7.08.15	7.08.15	HM (K)
2nd Lieut. Reginald Leak (20)	23.07.15	Chorlton-cum-Hardy	Rochdale	26.08.15	26.08.15	HM (D) w 7.08.15
2nd Lieut. Thomas Ralph Taylor (28)	5.05.15	Rochdale	Rochdale	7.08.15	7.08.15	HM (K)
2nd Lieut. George Cyril Thompson (24) (formerly 2/6th LF)	?	Nottingham?	Rochdale	24.10.15	24.10.15	LLC (D)
2nd Lieut. Godfrey Louis Wyatt (33)	5.05.15	London	Rochdale	24.10.15	24.05.15	LLC (D)

Total officers: 12

Helles Memorial. Photo Bill Sellars.

APPENDIX IX

Watkins & *The Gallipolian*

The Gallipoli Association was established in 1969 by veterans of the campaign. Their journal, The Gallipolian, *is still produced three times each year. Watkins gave the journal many extracts from* Lost Endeavour *as well as several articles, which are collected here.*

A Tree Grows in Brooklyn (and on Gallipoli)

ISSUE 6, 1971

On August 7th 1915, 2/Lt Eric Duckworth of No. 8 Platoon, 'B' Coy, 1/6th Bn Lancashire Fusiliers, went 'missing' near The Vineyard. We all know what 'missing' so often meant on Gallipoli.

In 1922 Eric's parents made a pilgrimage to the Peninsula where, in memory of their son they planted a small British Oak in Redoubt Cemetery, and where many of Eric's Platoon — all boys from Rochdale and district — found their last resting place. Eric's younger brother Geoffrey has also since made a pilgrimage to this sacred land of lost endeavour and where so many other trees and plants have been nurtured by the bones of other British and Colonial boys.

Geoffrey tells me this little oak sapling is now a big British Oak over 30 feet high. I like to think of this symbolic British Oak having weathered many a Mediterranean storm over the past 50 years, as we ourselves once weathered an equally violent storm there, in 1915.

Geoff is particularly appreciative of the kindness of Major Hughes, late of the Imperial War Graves Commission, who, after assisting to plant this little oak sapling, took water to it every day for a long time to foster its early growth. Such a kindness as this, as Geoff himself says, gives a nice warm glow to the heart. Me — I think of those two lines of the poem:

> Such things as this, and some more now and then
> Are part of the strong golden friendship of men.

The Gallipoli Membership List: An Appreciation

ISSUE 8, 1972

This detailed and accurate membership list (E & OE) of today's Gallipoli survivors — a prosaic bit of work, would you say? Or is it not in itself, perhaps, one of the most colourful bits of Gallipoli literature ever produced. Is it not fascinating beyond all telling to read through all these names, to read of the innumerable different Regiments and Ships, and let the imagination play on all their historic and colourful backgrounds. This scintillating panorama of Yeomanry and Infantry Regiments with their historic labels of bewildering variety, and all the famous Ships of the Royal Navy. It is really something. One gets reminded (at least I do) of the words of that famous marching song —

> See them flash by ... etc. etc.,
> ... they'll never die.
> For they'll live in song and story.

I always get reminded of this marching song whenever I read through this Membership List.

And best of all is the cosy brotherliness of it all. Lieut-Generals, Major-Generals, Admirals, Fleet Captains, Air Vice-Marshals, Lt-Colonels, Sqdn-Leaders — in fact all the illustrious Service ranks of our country and Commonwealth — they all jostle and mingle happily in this Membership List with Private Soldiers, Gunners, Signalmen, Troopers,

Leading Seamen, Ordinary Seamen, Sergts., Corpls., etc., etc. — all mingling together and with 'rank' all forgotten, in a vast sea of brotherhood. I think this is the nicest touch of all — this indiscriminate mingling of ranks one gets in the alphabetical list. And it gives, too, a panoramic slice of British and Commonwealth history — revealed by the names of all these regiments. Names like Australian Light Horse, Scottish Horse, the many different County Regiments of Yeomanry — each proudly and tenaciously retaining their own special and jealously guarded County titles — titles like Fife & Forfar Yeomanry; Queen Mary's Yeomanry; City of London Yeomanry — Rough Riders (I wonder how they got this latter title). All these Yeomanry and Horse Regiments, now dismounted and fighting on foot alongside the multifarious collection of equally historically named Infantry Regiments.

Names like Sherwood Foresters (Ah! Shades of Robin Hood); the many different Fusilier Regiments, including the Lancashires, of 'six before breakfast' fame. Then the many Scottish Regiments — the Lovat Scouts; The Black Watch, etc., etc. Impossible to list all these different regiments in this short appreciation — there were so many. And included, too, the tough regiments from 'that most distressful country' — and this time fighting alongside us and not agin us — as they like to at times — the Inniskilling Fusiliers, the Royal Munster Fusiliers; Connaught Rangers. All this colourful gang — the Tommies, the 'Jocks,' 'Hicks,' 'Taffys,' the 'Cobbers' and the 'Kiwis'. You'll need a whole library even to sketch the history of all these different regiments and their historic antecedents — all these chaps who have ever flocked to the Colours in times of our national danger. And not forgetting, too, the many different Territorial regiments of our own citizen soldiers, whose bones equally littered this tragic bit of land. And not forgetting, too, the strong silent types — those who wear the Navy Blue cloth — those who do so much and are so damned reticent of their own exploits — all this Navy Blue uniform and their many various ships — so much they did, and so little they said about it; the RNVR too.

Impossible, in this short appreciation to include *all* regiments and ships, but it would be a grievous omission not to mention specially the 'half and halves', the half-sailors and half-soldiers, and equally good at either job — the Marines. The Royal Naval Division boys, their battalions and their individual ships' names proclaiming their own naval origins: Drake, Hawke, Benbow, Howe, Collingwood, Hood and Anson battalions of the RND. And then the Portsmouth and Plymouth battalions of the old RMLI, now the Royal Marines. Stout-hearted, cheerful, able, these chaps, always on hand wherever the trouble was thickest. From the very start of the campaign in April and right to its bitter end in 1916, the RND was always on hand.

And, as if all this list of 'troops, assorted' were not enough, there yet remains to be mentioned the few RNAS machines and their crews, a few of the latter being members. How often, as a scruffy young soldier on Gallipoli, have I gazed wonderingly at these machines, gleaming in the bright sunlight and making a daily reconnaissance flight over the Peninsula, and with the Mediterranean blue sky peppered with the white puffs of bursting AA shells.

And to show up still more the versatility of the Gallipoli forces we find mention of officers who served with Gurkha and Sikh regiments.

All this fantastic and bewildering variety of warriors on this historic battleground, listed in our Membership List. What a goodly company of men to serve with. And what a goodly company of men to die with.

> He that outlives this day, and comes safe home
> Will stand tip-toe when this day is named.

Maybe Shakespeare was thinking of this gang when he wrote those lines.

And we that came safe home are doubly indebted to Major Banner for keeping alive this colourful scroll of survivors.

Brief Encounter

ISSUE 16, 1974

If Gallipoli had never happened I'd never have had the good fortune to be a member of this special Gallipoli Brotherhood of ours. And I'd never have met this tinker chap — this rabid lover of the English countryside — and of the Kentish countryside in particular. In my conceit — the insufferable conceit of the very young — I had thought I was the only person specially selected by God to really appreciate this Kentish countryside. That was, until I'd met this tinker chap.

I think it was Sir Walter Scott who best said it — about the love of one's own country:

> Breathes there a man with soul so dead
> Who never to himself hath said
> This is my home — my native land ... etc., etc.

This love of the countryside — it's something you've got — or you haven't got. If you've got it, it's a bit like first falling in love except that this love of the countryside persists with increasing intensity, making everything else seem of secondary importance. All Nature seems to combine to produce a masterpiece of perfect harmony — the sweet smell of the grass and flowers, the hum of the bees, the tinkling notes of hidden streams, bird songs, the incredible and artless beauty of the trees, the gentle breeze rustling the ripening corn fields into rolling waves of gold. All this, and more, I learned from this tinker. Not that he was very articulate — but some chaps don't seem to need the gift of words — he breathed and exuded his passionate love of England in every pore of his skin. I guess that some of you yourselves have met chaps like this — men who don't need to say a lot — or can't — but who in some mysterious way seem to convey everything they feel by some magic telepathy.

A simple, unassuming sort of bloke — this tinker chap. They're a dying race today — in fact probably quite extinct now, for who wants to have a kettle or pan mended, or scissors ground and sharpened when you can buy these darned things so cheaply in the stores. But years ago, tinkers were a fairly common sight — picturesque, untidy little men, fiercely independent, who carried on their trade quietly, giving offence to no one. Ex some Lancashire regiment, he'd served on Gallipoli until he bought one in the August offensive, but he stuck it out in other Middle East theatres of war until the end, when he was finally invalided out and with the usual miserly pension for his disability of a few shillings a week. Before the war he worked in some dim and unhealthy trade connected with tinning — close confined work in a badly lit and badly ventilated backroom. The war over, he decided not to go back to that life again. Like he said, 'Well mate, after all that fresh air and Middle East sunshine — well, y'know yerself, mate, how a man feels when he gets back to Blighty.' So he decided to roam around for a bit, finally coming down to Kent where he stayed, getting his living as a travelling tinker, mending kettles and pans, etc., a craft for which his previous trade training had given him a certain dexterity — canvassing the housewives in the villages in his own particular 'pitch'.

'Love every minute of it, I does, mate. England's a bloody good place to see when yer travels around the countryside — and all that — bloody lovely it is. Beats me why other folk don't wander round more and see it. I guess they must be bleedin' blind.'

I'm glad I met him. I met him quite by chance when I was living in Canterbury just after the First War. This tinker chap wasn't the only one who had fled, appalled from his previous occupation. The muck and grime of old Lancashire and after some four-and-a-half years of the Middle East, shocked me, so I moved on to Kent and settled in Canterbury.

Why Canterbury? Oh! It was just pure chance — a flying visit to an old wartime chum in Tunbridge Wells — an invitation to see a bit of Kent before settling down again in Lancashire. Not that I got much chance of seeing Kent in that short week-end, especially as it poured with rain the whole time. That didn't worry us too much — it's always fine weather in the Saloon Bar. But before leaving for home, he pressed me to have a look at Canterbury — told me I'd find it a bit different from looking at factory chimneys.

It was still raining hard when I arrived rather late in Canterbury in the evening, and I was fortunate enough to find good digs — just under the shadow of the ancient and historic Westgate Towers, but the next day and for the following fortnight the kindly gods — and with sly humour, must have conspired together to bring about my complete undoing — faultless sunny days, soft balmy nights heavy with the scent of blossom and the green, green grass of England. There's no scent in all the world like it.

Those of you who know Kent need no eulogy from me to extol its unbelievable charm just after the First War — and especially to an ex-Lancashire mill-hand. I couldn't believe it was real. All counties have their own especial appeal but it so happened I had stumbled on the

Kentish countryside — and Canterbury — about Maytime — the blossom-laden orchards, the bronzed faced land girls training the hop vines around the hop poles — oh! there's no end to the catalogue of enticing charms of this countryside. For me it was a case of total seduction by this corner of Paradise and I never tired of exploring its charms. It was one evening later in June that I'd met this tinker chap just on the outskirts of the city, attracted by the makeshift little tent and the glow of his little bivouac fire. There's no need for me to tell any other old soldier how a little camp fire pulls you towards it like a magnet — or maybe it's a hangover from the cave days. So I stopped and we chatted. Those were the days — before TV and radio had battered us into a state of moronic imbecility — they were the days when men found their pleasure in talking to their fellow men.

So we talked, and smoked, and talked and talked till the stars came out and all the time his wife — or woman — brooded anxiously over him like a hen over a sick chick — especially when he got an extra bad fit of coughing. It seems a bit of Turkish shrapnel had done his lung a bit of no good but he shrugged it off as of no importance — this persistent cough of his.

In an indiscreet moment I remember asking him didn't he want something better than what he was doing, but his reply silenced me. It was one of those soft balmy June nights where everything seemed to hang quivering halfway between heaven and earth — the sort of nights you only get in an English summer. The chimes from the distant Cathedral Bell Harry tower echoed on the night air, soft as a Gregorian chant, and emphasising still more the ecclesiastical tone of this ancient and beautiful city of Canterbury. He drew a deep sigh and after a time said solemnly, 'Can there be owt better in the world that this?' Then he gave a merry laugh and added, 'I reckon God Himself spends his annual holidays round these parts.'

So I 'hushed my big mouf' and listened while he talked, and talked. All about his every day roaming the lanes and by-ways, soaking up the beauty of England like a sponge. Occasionally his talking would be interrupted by that goddamned cough of his and the woman moved to put some more wood on the fire. 'What the hell did you do to him out there?' she said to me angrily. 'A right strong young fellah he was afore he went out there, and now…'

And she gave a deep sigh that seemed to come up right from the bottom of her battered and worn shoes.

'Wanted to live all his days' the tinker told me, 'doing nothing else but what he was doing now.' And when he died, he said, he wanted to die — not in a hospital bed, but where he could see his beloved land for the last time. I try to recapture his sentiments in the following poem — and may God and my fellow Gallipolians forgive me!

> Just take off my coat and fold it
> And prop it high under my head
> And sit by me there, and wi' your hands in mine,
> Just sit by me there... till I'm dead.
> Nay, don't fret yerself, lass,
> And don't pipe yer eye.
> This way I wanted to die.
>
> A'top of this hill, and under a tree,
> Where all around me I can see
> This dear, dear land of England.
> The hedges an' farms and fields by the score
> An' the long winding roads, an' the woods and the streams,
> The parks an' the manors — the Gentry's abodes.
> An' see the white smoke curl up from cottage and inn...
> The littler dwellings o' men and their kin.
> I've roamed every by-way and lane to the end
> Wi' my scissors to sharpen and kettles to mend.
> ... All this 'as bin my 'pitch' ... as far as yer eye can see,
> An' belongs only to England ... an' me.
>
> What! crying, lass? Nay — don't weaken like that.
> It's bin a good life ... an' I've bin kinda lucky wi' you as a wife.
> I've 'ad ups and downs ... and a good share of fun,
> Tho' I've done many things I didn't oughter 'ave done.
> But I've learnt on my 'pitch' all the Rules of the Road,
> Like ... 'elping a bloke when he's down.
> An' learning to carry my share of the load.
> ... that a frown and a scowl ain't worth not the half of the worth
> of a smile In helping a bloke to do the Last Mile.

An' I've learnt that these things, an' some more now an' then,
Are part o' the strong golden friendship of men.

It ain't much, I agree, but it'll 'elp me, mebbe
When I goes up aloft and I stands before ... Him.
... When I stands, cap in hand ... an' a shufflin' my feet ... and remembring my sins,
An' his welcome won't seem half so grim.
He'll frown ... maybe clout me a bit ...
Then He'll gimme His hand.
And say 'Damme, yer naught but a tinker,
Yer didn't understand.'

The 1976 Annual Dinner: Idle Thoughts by an Idle Fellow

ISSUE 22, 1976

Grow old with me
The best is yet to be

With the ever-increasing pains and penalties of the ever encroaching years, I would recall these poetic lines with increasing sour disapproval, I even got some satisfaction from contemplating the pleasure it would give me — just once to meet this poet and kick him in the place where it hurts most. But that was before I became a member of the Gallipoli Association and attended our Annual Dinners.

Old age at its best — these functions. Old age at its noblest, its most serene, its most sensible, its most mellow... and above all, its most inspiring. Gatherings of much be-medalled grey-haired old warriors, supremely happy in a reunion of old pals. Remembrance of old times, old times so often crammed with memories of much suffering... and much glory. Remembrance of a now almost forgotten campaign, but a campaign unique in British history — Gallipoli. From the pit of this campaign shines like a beacon the heroism and comradeship of the multi-regiment force, striving amid the welter of so many mistakes and mismanagement, striving to the utmost for ultimate victory, only to be cheated at last of what was so very nearly in their grasp. Heroism of the highest order, comradeship never again to be equalled. The heroism is now past and becomes part of our nation's history. The comradeship lives on undiminished, and will remain undiminished until the last Gallipoli survivor has departed for ever, leaving only the fragrant memory of the soldier and sailor at his best.

All these thoughts crowded me at this year's Annual Gallipoli Association Dinner — this gathering of grey-haired survivors — this most exclusive club in all the world. The average age of the members must be well into the 80s. Our President, Lt-General Sir Reginald Savory, KCIE CB DSO MC, humorously rejects the label 'old age' — insists, as his gaze wanders over the spritely and active survivors attending this Dinner — insists it is not 'old age' but 'late middle age.' Sir Reggie, as always, addressed the meeting in his own inimitable and pleasing style — warm, cosy and honest. Despite his title and high military rank, he prefers to be simply 'Reggie' to all of us. No snobbery, no 'talking-down' to any of us — not with this bloke. I rate him as one of the country's truest democrats.

> The rank is but the guinea stamp
> The man's the man for a' that

Methinks this must have been the first poem that Reggie lisped at his mother's knee. His obvious 'matey-ness' reflects his attitude to his fellow men and comes over strongly as he addresses us. Best of all, I like the way he presides at these Dinners, his unique 'warmth.' Like a bland and benign Buddha he dominates the scene, his mild and kindly manner giving the lie to his aggressive square-jawed Churchillian image. His measured slow words row us along a pleasant stream of memories of a campaign of more than 60 years ago, the like of which has never illumined the pages of history more vividly.

My only grouse about these Dinners is that they always make me, with my mere trio of 'Pip, Squeak and Wilfred' medals, they make me feel naked. They seem lost, these three little medals of mine, among the vast array of the numerous medals and decorations displayed across the manly breasts of most of the other survivors gathered here. Some of them have so many medals, Orders and other decorations that, if in full military regalia — and if you dropped any of them in a swimming bath, they'd sink to the bottom thro' sheer weight of metal.

And these sort of medals they're wearing — well, you don't get them with Green Shield Stamps!!

His speech finished, Sir Reggie introduced our Guest of Honour, Field Marshal Sir Gerald Templer, KG GCB GCMG DSO.

We have indeed been honoured by his presence. And he had, like our own President, a nice easy way of talking, modest and quiet — just a soldier talking to other soldiers. He especially 'rang a bell' with me when he mentioned that at times he had been 'terrified' during his service in the trenches in France. (Up to now I had thought I was the only one who would admit to having been 'terrified'). But best of all I liked his modesty — (and, God knows, these high-ranking blokes are nothing if not modest) but even his modesty cannot hide the colourful background of his highly successful Service career. He recounted a particularly heart-warming story of an occasion when he and his wife paid a visit to our 'blessed' Peninsula. But it is the Field Marshal's own story and I wouldn't spoil it by re-telling it myself.

We had some nice appreciative words, too, from the Navy — Captain Eric Bush, DSO**, DSC, Royal Navy. No gathering is complete without the presence of this sparsely-built, much decorated Naval Captain, this precocious youngster, who as a 'snottie' at the tender age of 15, won his first decoration at Gallipoli. And judging by the crowded line of medals across his chest, it seems as if he made a hobby too of collecting umpteen other decorations and honours — in between his Naval duties, ranging from Midshipman to commanding ships of the Royal Navy. His own book *Gallipoli* is a monumental and very readable book on the Gallipoli Campaign and will rank as a landmark in military literature.

As the evening progressed, other members were invited to recount some of their own experiences. We would have gone on for hours and hours. One of the many blessings of the fermented juice of the grape is that it loosens the tongue, and, which is a good thing too, at these Dinners. For old soldiers and sailors — they all have one great defect — they all tend to be too damned reticent. Or is it that they all subscribe to that passage in Shakespeare's *Henry V.*

> In peace, there's nothing so becomes a man
> As a modest stillness and humility.

Tongue-tied, modest and reticent — that's the British soldier and sailor. May they never change. Last but not least, our Chairman and Hon Secretary, Edgar Banner, the 'indefatigable Edgar,' without whose boundless devotion, energy, expertise and finesse these functions would not so successfully exist. May Allah preserve him for many more years to come.

Our Ghostly Guests

ISSUE 25, 1977

They're with us at every Annual Dinner. Even if we cannot see them, we know they are there. No places have been laid for them, no Menu Cards, no numbered places to indicate their position 'above or below the salt.' No name cards for them. In fact, no nothing.

But, as one of them was heard to remark to his pal — 'Thank goodness, we, at any rate don't need all this paraphernalia. We just crowd in and are happy being with them once again.'

They are a privileged crowd, these disembodied member-guests of ours. Even prohibitive rail fares don't deter them from travelling — like it does so often with we earth-bound mortals. Even the Ticket Collectors on the trains turn a blind eye to their presence — pretend not to see them, and don't even bother to punch their tickets.

There's a particular one of these ghostly guests I always cotton-on to. I have a special affection for him. One of the earliest casualties of the Dardanelles campaign, unfortunately, this particular young fellah — and one of our best poets too. I think I like him most because of his white-hot love of his native land, and the way he reflects it in almost every line of his poems. And in the special way of his own words as he talks of his beloved England.

His name? Well, these disembodied guests have a 'thing' about name dropping. They don't allow it. But maybe we'll be able to identify him.

APPENDIX IX

If we can't, then more shame to us. This is the chap who reflects in his verses some of the many qualities of his own beloved country.

> Oh! there the chestnuts, summer through
> Beside the river make for you
> A tunnel of green gloom...
> The stream mysterious glides beneath.
> Oh! damn! I know it! and I know
> How the May fields all golden glow...
> the dews are soft beneath a morn of gold
> For England's the one land I know
> Where men with splendid Hearts may go...
> There's peace and holy quiet there.
> Great clouds along pacific skies
> And men and women with straight eyes,
> Lithe children, lovelier than a dream,
> A bosky wood, a slumbrous stream...
> And often, ere the night is born,
> Do hares come out about the corn?
> Oh! is the water sweet and cool,
> Gentle and brown, above the pool?
> Say, is there Beauty yet to find?
> And Certainty? and Quiet kind?
> Deep meadows yet, for to forget
> The lies, and truths and pain? ... Oh! yet
> Stands the Church clock at ten to three?
> And is there honey still for tea?

No need, I think, to rack our brains to identify this particular Ghostly Guest.

THE GALLIPOLI ASSOCIATION
gallipoli-association.org

MAPS

Map 1   Rochdale and environs
Map 2   Gallipoli
Map 3   Helles
Map 4   Morning of 4 June 1915
Map 5   Evening of 4 June 1915
Map 6   Morning of 7 August 1915 — before the advance
Map 7   Afternoon of 7 August 1915 — after the advance
Map 8   42nd Division section from 19 August 1915

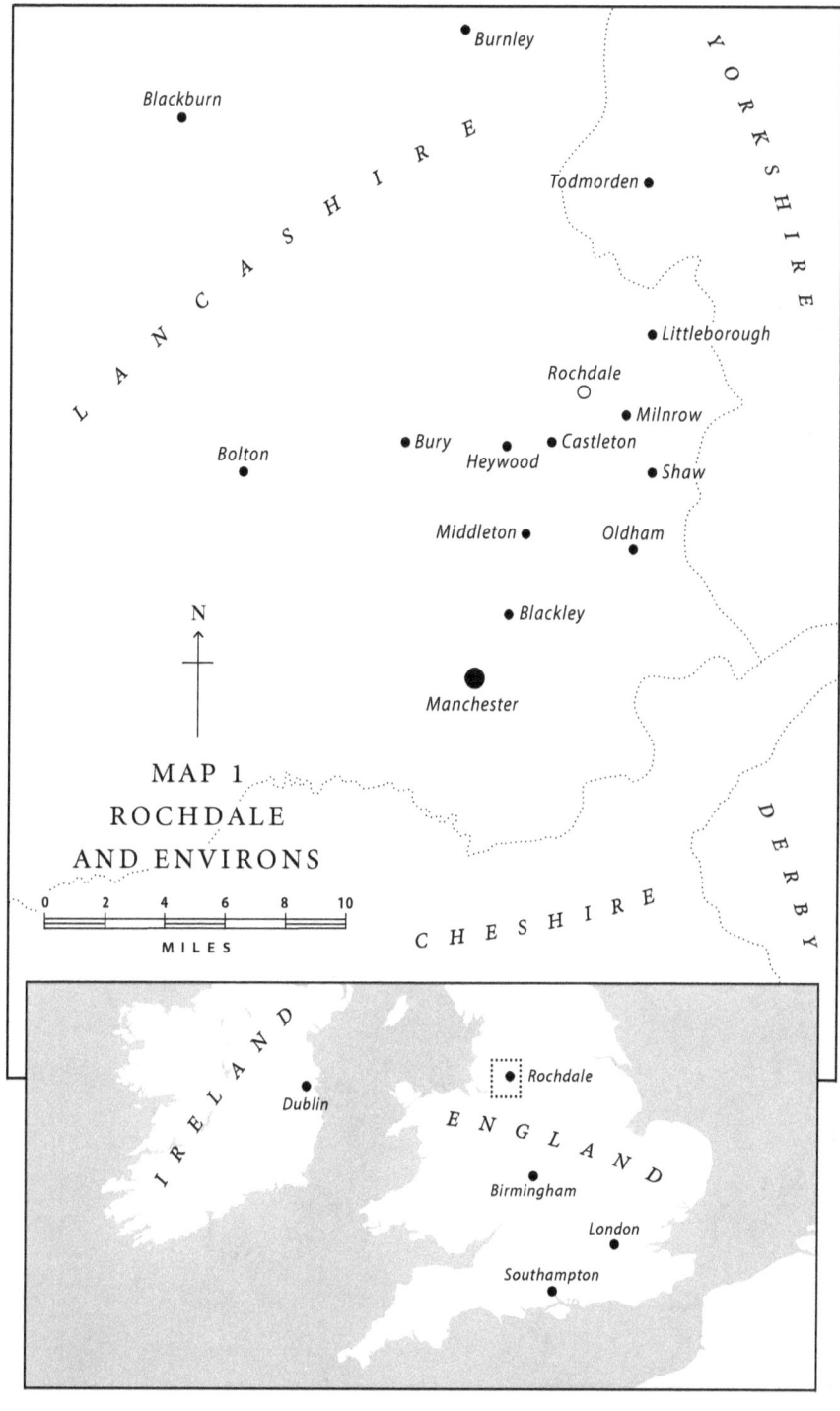

MAP 1
ROCHDALE
AND ENVIRONS

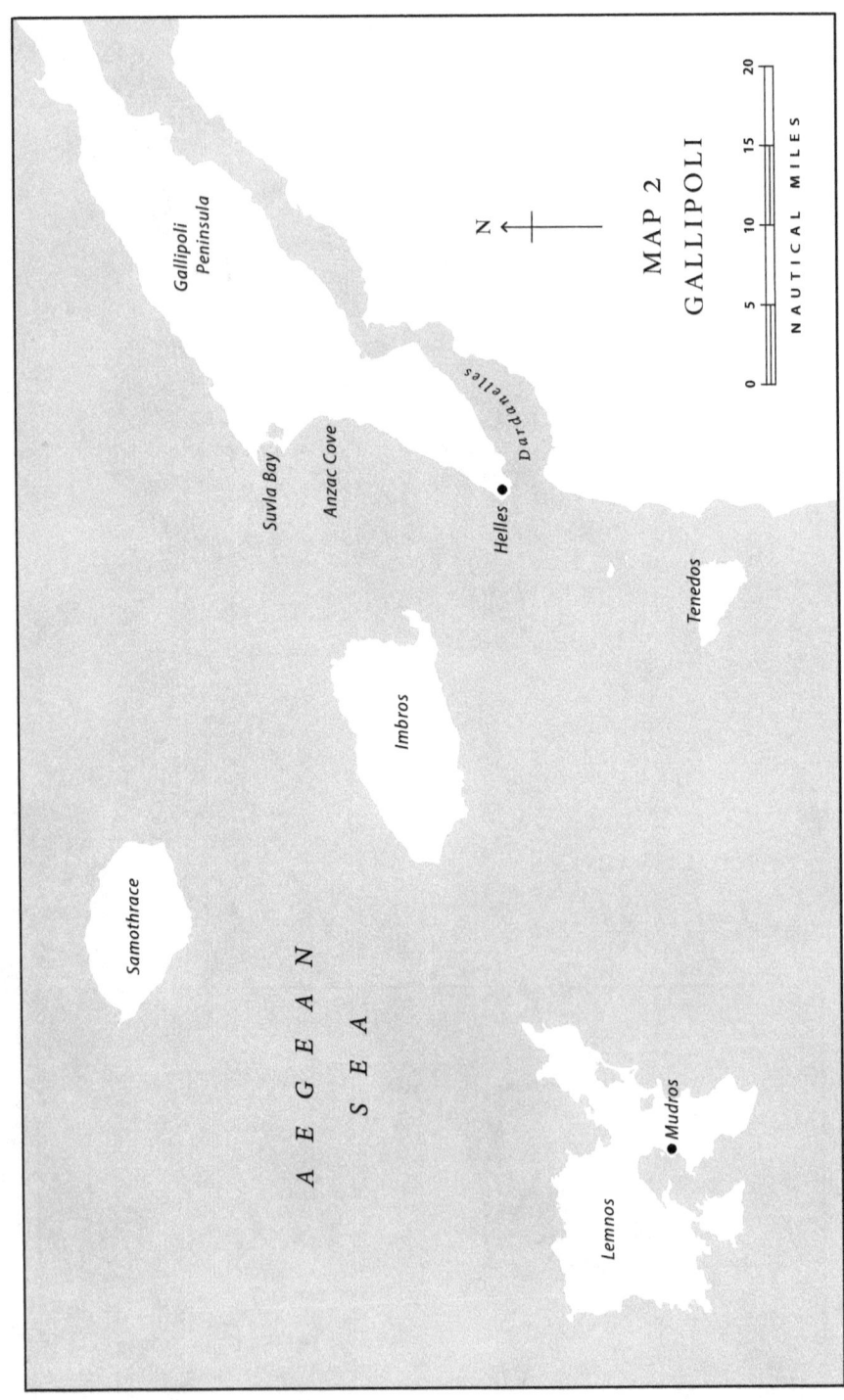

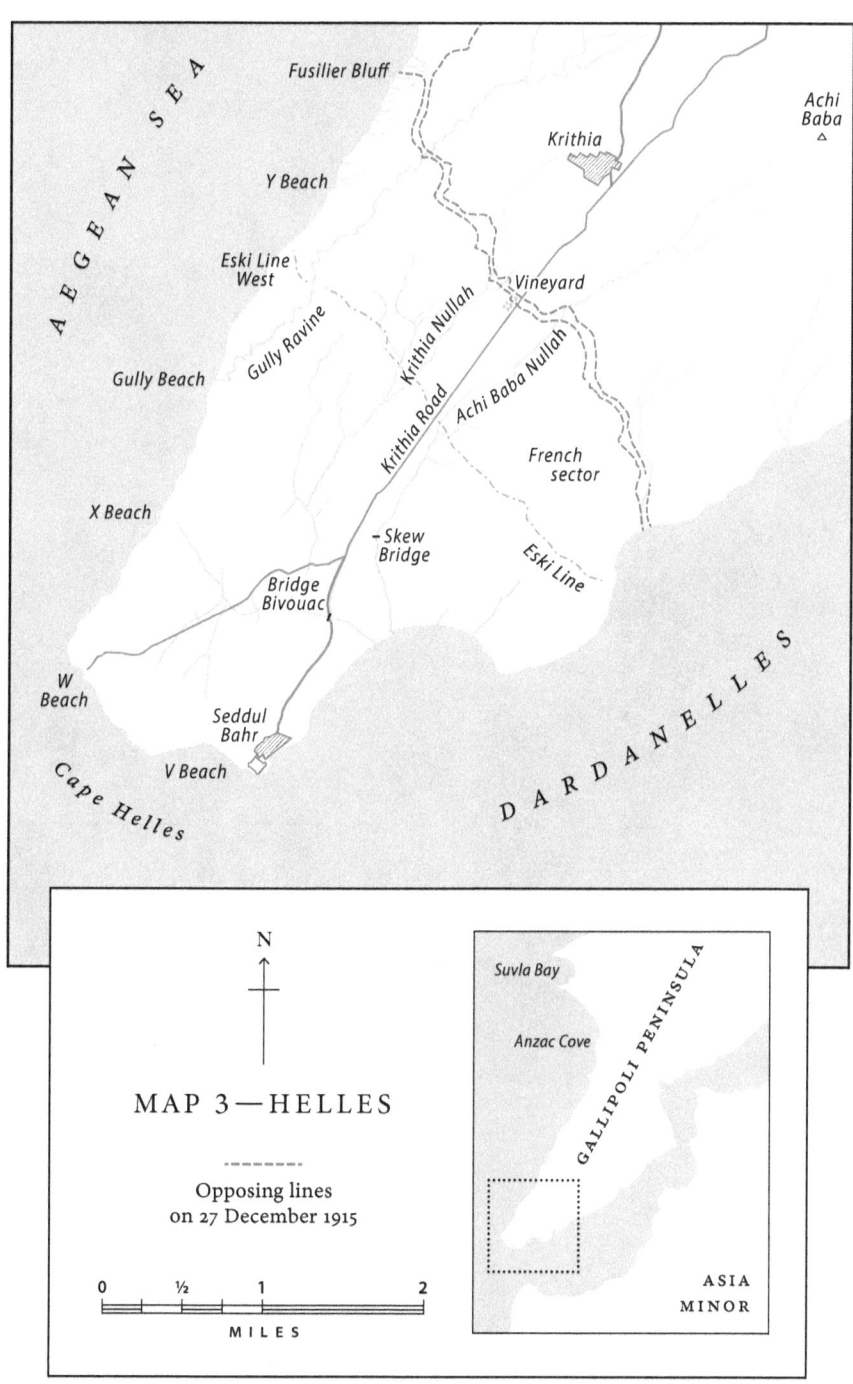

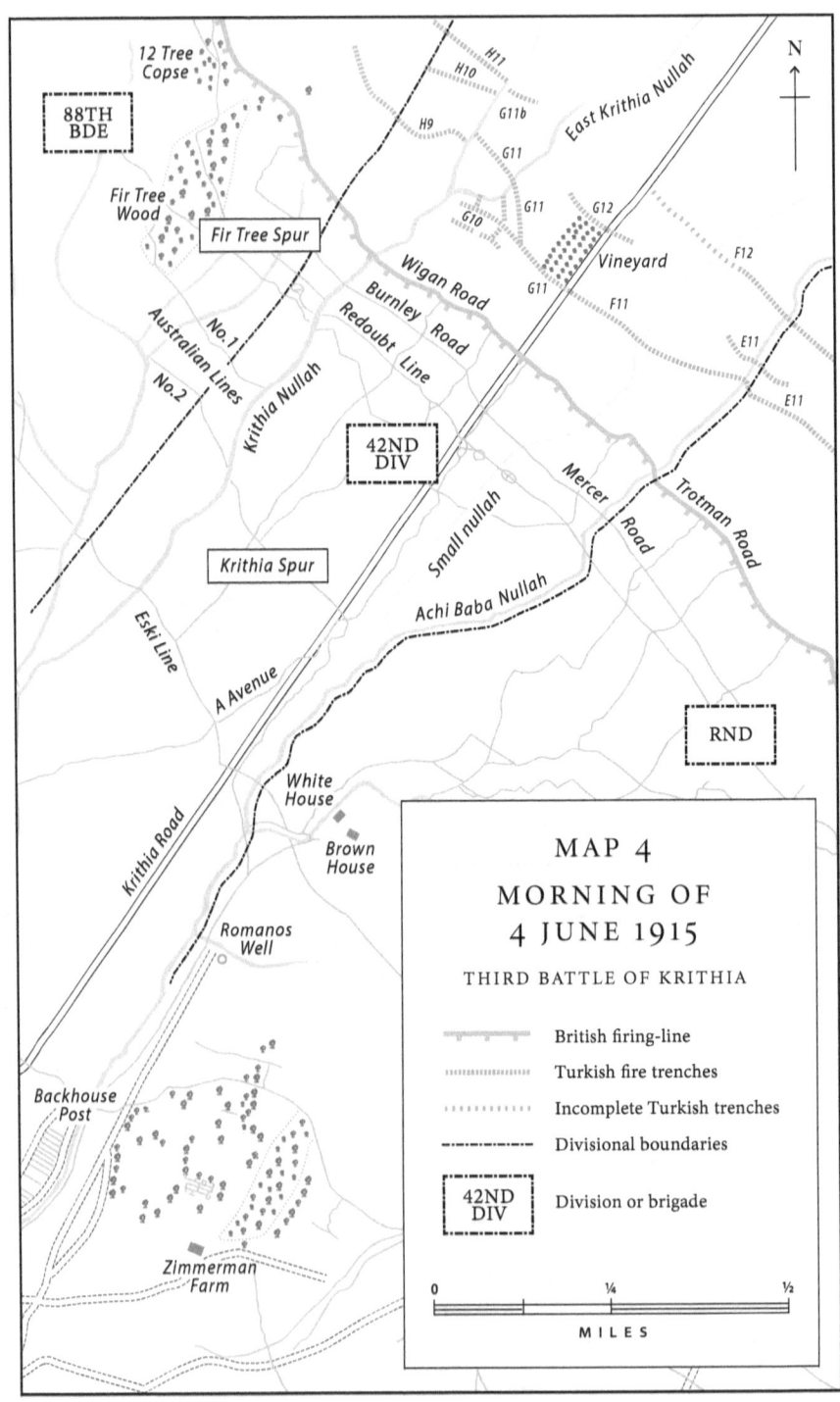

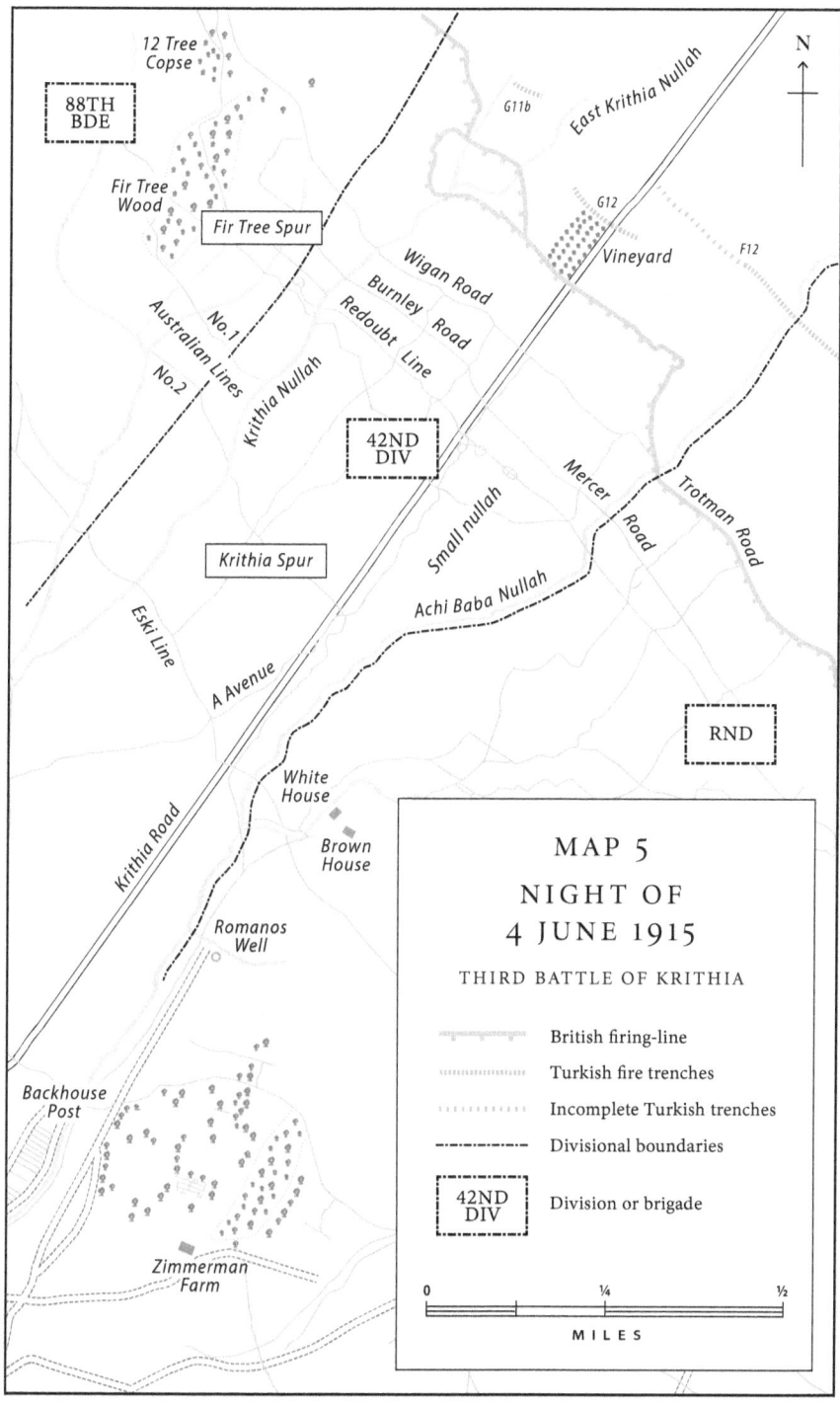

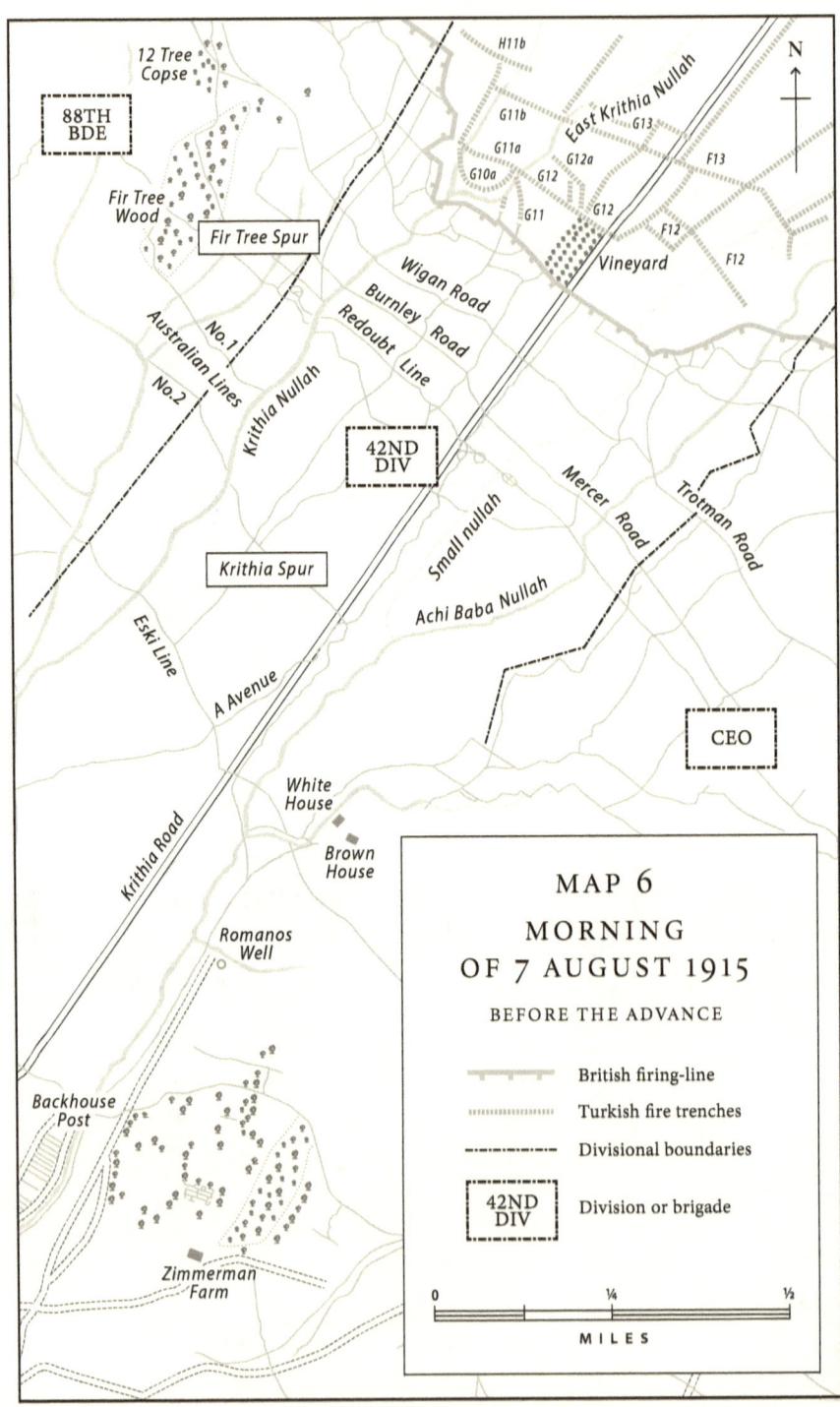

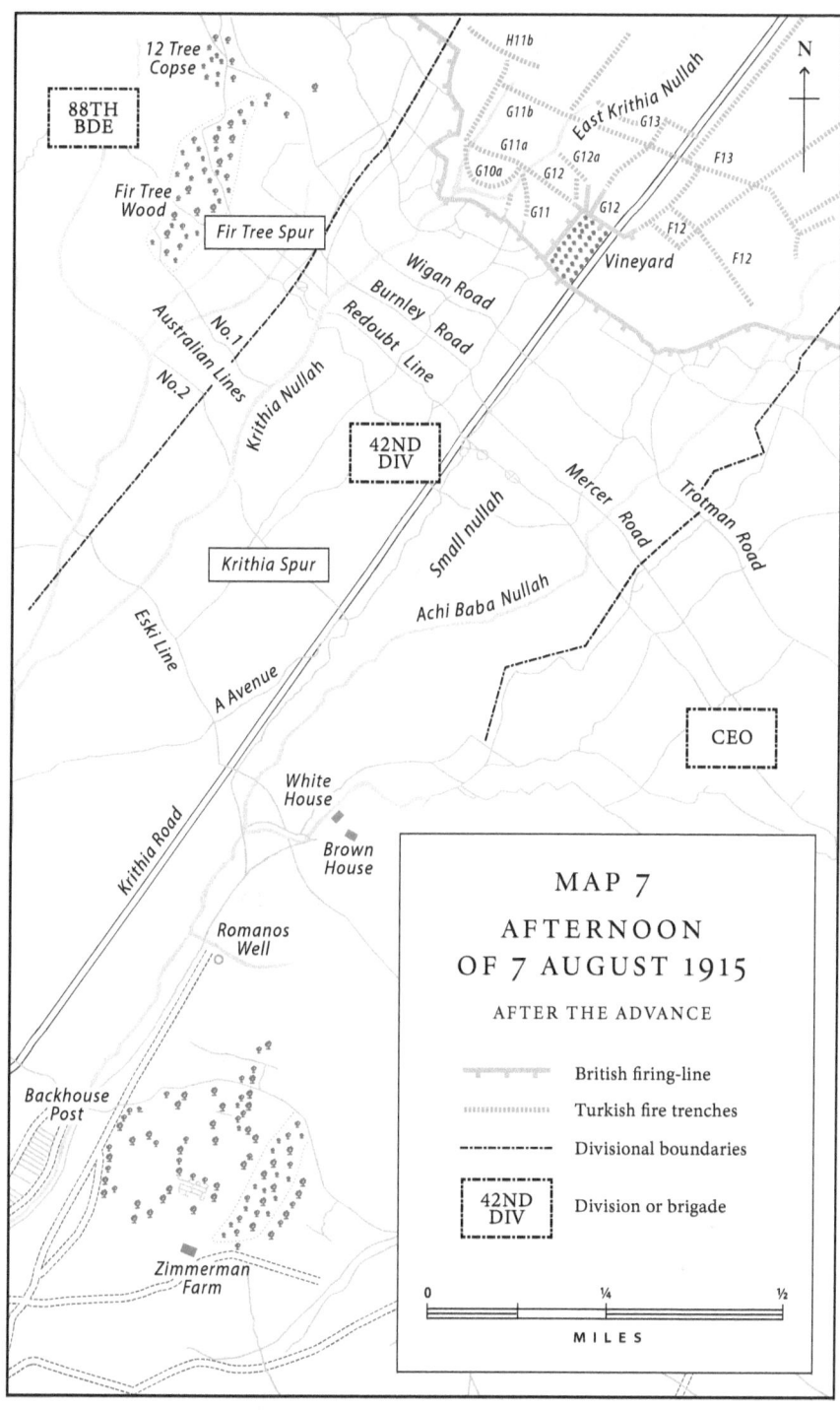

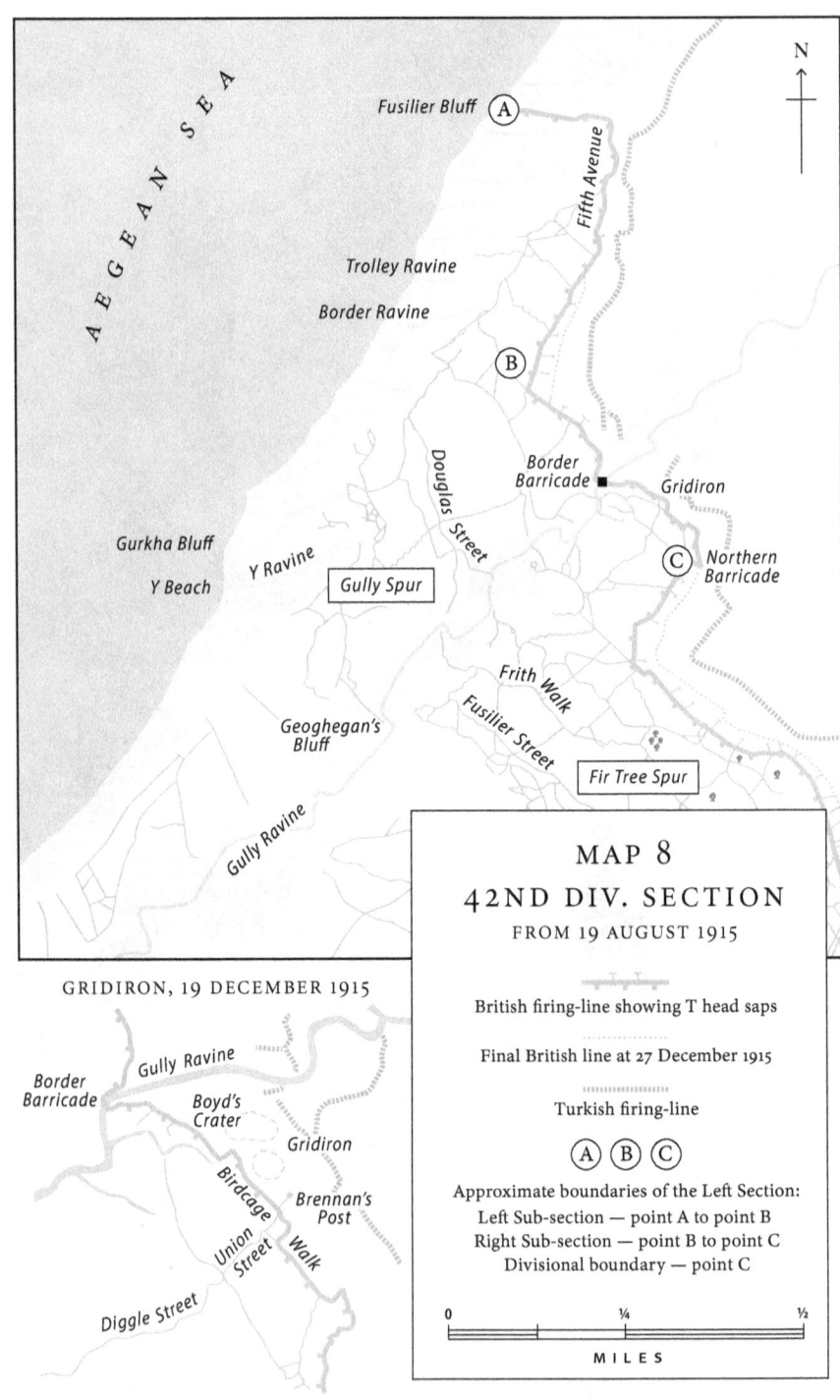

ABBREVIATIONS AND ACRONYMS

1/AM	Air Mechanic First Class
2/AM	Air Mechanic Second Class
2ic	Second-in-command
A/	Acting
AA & QMG	Assistant Adjutant and Quartermaster General
AB 64	Army Book 64, army pay book
ADC	Aide-de-Camp
ADMS	Assistant Director of Medical Services
ADVS	Assistant Director of Veterinary Services
AIF	Australian Imperial Force
APM	Assistant Provost Marshal
ASC	Army Service Corps
Bde	Brigade
Capt.	Captain
CB	Confined to Barracks, or Companion of the Order of the Bath
CE, C of E	Church of England
CEO	Corps expéditionnaire d'Orient
CMG	Companion of the Order of St Michael and St George
CO	Commanding Officer
Coy	Company
Cpl	Corporal
CQMS	Company Quartermaster Sergeant

CRE	Commander Royal Engineers
CWGC	Commonwealth War Graves Commission
DAA & QMG	Deputy Assistant Adjutant and Quartermaster General
DADMS	Deputy Assistant Director Medical Services
DADOS	Deputy Assistant Director Ordnance Services
DAQMG	Deputy Assistant Quartermaster General
DCM	Distinguished Conduct Medal
DL	Deputy Lieutenant, honourary position, typically of a county
DSC	Distinguished Service Cross
DSO	Distinguished Service Order (asterisk indicates a subsequent award, worn as a bar on the ribbon of the original award)
ERA Petty Officer	Engine room artificer, with non-commissioned rank of Petty Officer
GCB	Knight Grand Cross of the Order of the Bath
GCMG	Knight Grand Cross of the Order of St Michael and St George
GOC	General Officer Commanding
GSO	General Staff Officer
HM Forces	His Majesty's Forces
HQ	Headquarters
KBS	Kite Balloon Section
KCIE	Knight Commander of the Order of the Indian Empire
KG	Knight Companion of the Order of the Garter
KIA	Killed in Action
KOSB	Kings Own Scottish Borderers
L/Cpl	Lance Corporal
LFs	Lancashire Fusiliers
LG	London Gazette

Lieut., Lt	Lieutenant
MBE	Member of the Order of the British Empire
MC	Military Cross
MG	Machine-gun
MO	Medical Officer
MP	Military Policeman, or Member of Parliament
MR	Manchester Regiment
NCO	Non-Commissioned Officer
OBE	Order of the British Empire
OC	Officer Commanding
OD	Other Denominations
Pte	Private
QM	Quartermaster
QMS	Quartermaster Sergeant
RAF	Royal Air Force
RAMC	Royal Army Medical Corps
RC	Roman Catholic
RE	Royal Engineers
Regt	Regiment
RFA	Royal Field Artillery
RFC	Royal Flying Corps
RN	Royal Navy
RND	Royal Naval Division
RSM	Regimental Sergeant Major
Sgt	Sergeant
St	Saint
TD	Territorial Decoration
VC	Victoria Cross
VD	Volunteer Officers' Decoration
VO	Veterinary Officer
YMCA	The Young Men's Christian Association

BIBLIOGRAPHY

Archival sources

Correspondence between Charles Watkins and Colonel Carl K. Mahakian, 1985 (editors' collection)

Charles Watkins, transcript of interview with Peter Liddle in February 1973, Liddle Collection, Leeds University Library, LIDDLE/WW1/GALL/REC/223

War diaries, 1/5th, 1/6th, 1/7th and 1/8th Lancashire Fusiliers, 125th Infantry Brigade HQ, and 42nd Division, General Staff & 'A' Section, The National Archives (series WO 95)

British Army service records, pension records and medal index cards, The National Archives (series WO 363, WO 364, WO 372)

Royal Air Force service records, The National Archives (series AIR 76 and AIR 79)

Published sources

Soldiers Died in the Great War, 1914–1919 (His Majesty's Stationery Office, 1921)

The Gallipolian, journal of the Gallipoli Association, gallipoli-association.org

Bigwood, George, *The Lancashire Fighting Territorials* (London: 'Country Life' & George Newnes, 1916)

Crane, Michael and Bernard de Broglio, 'Sorry, you two Australian Mums', *Digger*, Magazine of the Families and Friends of the First AIF, no. 78, March 2022

Gibbon, Frederick P., *The 42nd (East Lancashire) Division: 1914–1918* (London: Country Life, 1920)

Jones, H.A., *The War in the Air: Being the Story of the Part Played in the Great War by the Royal Air Force Vol. VI* (Oxford: Clarendon Press, 1937)

Laffin, John, *Damn the Dardanelles!* (London: Osprey, 1980)

Purdy, Martin, and Ian Dawson, *The Gallipoli Oak* (Ramsbottom, Lancashire: Moonraker Publishing, 2013)

Riley, Alec, edited by Michael Crane, Bernard de Broglio, *Egypt Diary 1914–1915* (Mosman, NSW: Little Gully Publishing, 2022)

Riley, Alec, edited by Michael Crane, Bernard de Broglio, *Gallipoli Diary 1915* (Mosman, NSW: Little Gully Publishing, 2021)

Westlake, Ray, *British Regiments at Gallipoli* (Barnsley, South Yorkshire: Pen & Sword, 2004)

Westlake, Ray, *The Territorials 1908–1914* (Barnsley, South Yorkshire: Pen & Sword, 2011)

EDITORS' ACKNOWLEDGEMENTS

Mike and Bern thank friends and family for their help in producing this book. Bill Sellars, our man in Eceabat, gave us boots on the ground, as well as invaluable editorial advice and help with fact checking. Mike's sister Norma Wallworth researched the family histories of Charles Watkins and Henry William Griffiths. Reinhard W. Filla scrutinised the maps with a critical eye.

The editors thank Philip Mather and his team at the Lancashire Fusiliers Museum in Bury for access to their archive, and for two images reproduced in this book: the portrait of Lord Rochdale (p. 232) and the group photograph of officers of the 6th Lancs Fusiliers in Egypt (p. 286).

The authors of *The Gallipoli Oak*, Martin Purdy and Ian Dawson, were generous in providing us with the photograph of No. 8 Platoon's reunion in 1938 that features Charles Watkins (p. 214).

The Gallipoli Association, of which both editors are proud members, kindly permitted us to reproduce the articles by Watkins that were published in their journal.

We are similarly grateful to Stephen Chambers, Jim Grundy, Peter Hart, Michael Robson and Warren Smith for their continued support of Little Gully's endeavours.

Last but definitely not least, Mike and Bern would like to thank their respective spouses: Viv Lacey for her editorial assistance and Cheryl Ward for her artwork, and both for their continued toleration of two Gallipoli Campaign obsessives.

INDEX

29th Division 33, 273, 275, 276, 282
  1st KOSBs 253, 274, 275
  1st Lancs Fusiliers 3, 26, 26–27, 36
  1st Royal Inniskilling Fusiliers
    275, 308
  1st Royal Munster Fusiliers 308
  4th Worcester Regt 282
  88th Brigade 251, 274, 279,
    324–327
42nd (East Lancashire) Division viii,
    42, 216, 245, 251, 258–262,
    276, 281–283, 324–328
  South Eastern Mounted Brigade
    285
  125th (Lancs Fusiliers) Brigade 230,
    251, 275, 276, 284, 285
  1/5th Lancs Fusiliers 25, 69, 260,
    263, 273, 280, 283, 284
  1/6th Lancs Fusiliers 24–25, 215,
    216, *217*, *218*, 219, 220, 227,
    233, 238, 243, 248, 251–252,
    253, 254, 256, 260, *286*. See
    also Lord Rochdale
  No. 8 Platoon *214*, 306, 334. See
    also Platoon Sergeant
  A Company 282, 283
  B Company 282, 283
  C Company 268, 283
  D Company 125, 268, 283

Gallipoli establishment, draft,
    casualties 287–289
Gallipoli narrative 263–271
Gallipoli Roll of Honour
    290–304
Gallipoli timeline 273–285
  1/7th Lancs Fusiliers 32, 251, 254,
    260, 269, 270, 273, 274, 275,
    281, 283, 284, 285
  1/8th Lancs Fusiliers 260, 273, 278,
    280, 283, 284, 285
  126th (East Lancs) Brigade 233, 251,
    254, 276, 277, 278, 281, 283
  1/4th East Lancs Regt 260, 281,
    282, 283
  1/5th East Lancs Regt 260
  1/9th Manchester Regt 251, 258,
    260, 280
  1/10th Manchester Regt 260
  127th (Manchester) Brigade 233,
    251, 268, 269, 275, 276, 277,
    279, 283
  1/5th Manchester Regt 251, 261,
    279
  1/6th Manchester Regt 25,
    165–166, 250–252, 261
  1/7th Manchester Regt 261, 270
  1/8th Manchester Regt 252, 261
Composite Division 275

INDEX

A

A Avenue 276, *324–327*
Achi Baba 93, 157, 250, 264, 268, 271, 274, *323*
Achi Baba Nullah 243, 277, 278, 279, *323–327*
aircraft 158, 309
Alexander, Lieut. Col. C.T. 284
Alexandria 60, 85, 191, 192, 199, 219, 238, 273
Anzacs vii, 23, 74, 143–153, 219, 242–244, 252–253
  Australian kit 49, 50, 51, 76
army pay book 102–103, 104
Asiatic Annie 271
Australian Line, nos. 1 and 2 276, 279, 282, *324–327*

B

Balkans. *See* Macedonia
Banner, Major [E.H.W.] 309, 317
Barker, Capt. R. 268, *286*
Bartley, Lieut. T.D.M. 268, *286*
battles
  Action of Achi Baba Nullah 278
  Battle of 6/7 August – Battle of the Vineyard 25, 157, 158, 160, 165, 171, 248–249, 250–252, 279–282, *326–327*
  Diversionary actions, 19 Dec 1915 114, 240, 254, 285, 289
  Second Battle of Krithia 220, 236, 240, 253, 274–275, 288
  Third Battle of Krithia 111–115, 268, 277, 288
Berrington, Lieut. J.S.D. 268
Bigwood, George 263
Birdcage Walk 254, 256, 285, *328*
Bletsoe 52
Border Barricade 253, 254, 256, 282, 283, *328*
Boyd, Capt. 254
Boyd's Crater 254, *328*
Brennan's Post 285, *328*
Brentnall, Lieut. 286, 287
Bridge Bivouac 253, 275, 276, *323*
Bright, John 24
Buckley, Fred [William?] 108–110, 238–239
Burgess, Sgt James 236
burials 87, 121, 150–153, 199, 200, 201, 256
Burnley Road 276–277, *324–327*
Bush, Capt. E.W. 3, 316

C

Cairo 23, 29, 69, 83–86, 161–162, 245
  Citadel 29, 30, 60, 139–140
  Mena Camp 145
  prostitution 84–86, 203
Carton de Wiart, Lieut. M. 268, *286*, 288
cemeteries. *See* burials
chaplains 104–109, 148–149, 202, 269
  Fletcher, Rev. Denis 269
Christmas 93, 190, 202–203, 204, 255–256, 269–270
Clegg, Capt. A.V. 286, 289, 304
Coe, 2/Lt T.H. 287
Cooke, 2/Lt Charles Earsham 251, 280
cotton mills 22, 24, 68–69, 123–124, 145, 183, 219
  Sparth Mill 122
  Standard Mill 123
  Turner Brothers 215
Cpl I—— 57–64, 65, 67, 236
Crossley, Capt. 286
Cryer, Pte John 251

D

Davies, Cpl Chadwick 240, 294
Davies, Lieut. P.V. 268, *286*, 288

INDEX

de Robeck, Vice Admiral 26
Diggle Street 284, *328*
discipline, military 61–64, 109, 145–148, 194, 205
Douglas, Maj. Gen. William 251–252, 258, 263
Douglas Street 283, 284, *328*
drafts. *See* reinforcements
Duckworth, 2/Lt Eric 214, 227, *286*, 289, 304, 306–307
  Mary Duckworth *214*
Duckworth, Pte Samuel 238, 294

E

East Krithia Nullah. *See* Krithia Nullah
equipment 50–51, *217*, *218*. *See also* Australian kit, machine-guns
  oilskins 204–205
Ermine, SS 285
Eski Line 256, 270, 276, 284, 285, *323–327*
evacuation viii, 11, 12, 203, 204, 205–209, 210–212, 256–257, 270, 285. *See also* Diversionary actions, 19 Dec 1915
  self-firing rifles 206

F

Felstead, Bombardier William Herbert 242–244
Fifth Avenue 253, 254, *328*
Fir Tree Spur 251, *324–328*
Fitzgerald, Edmund 52
flies 175–176
flowers 96–98
food and drink 16–17, 19, 51–52, 56–57, 70, 74, 82, 155, 157, 158, 160, 167, 175–176, 182, 191, 218, 265. *See also* tobacco
  beer 39, 49, 58, 64, 73, 120
  chocolate 118
  emergency rations 66
  jam 127–134, 167, 175, 181
  lime juice 125
  officers' rations 68, 170
  rum 34, 51–52, 57, 71, 84, 89, 112, 117, 182, 183, 192–195, 196, 201–202
Forshaw, Capt. William, VC 251
French expeditionary force (CEO) 7, 44, 69, 266, 278, *326–327*
Frith, Brig. Gen. Herbert Cockayne 15–22, *228*, *229–230*, 260
Frith Walk 284, *328*
Fusilier Bluff 238, 253, 254, 282, *323*, *328*
Fusilier Street 283, 284, *328*

G

Gallipoli Association 224–225, 227, 306–309, 314–317, *319*
Geoghegan's Bluff 282, 284, *328*
George, philosopher 119–120, 181, 182–183
Gibney [McGibney?], Jack 30–32, 235
Gledhill, Capt. J.J. 268, *286*, 288
Greenwood, L/Cpl 11, 120
Gridiron, the 240, 254, 285, 289, *328*
Griffiths, Capt. William Henry 11, 170–174, 245–249, *249*, 280, *286*, 289, 304
Grimes, Cpl 78, 83–87, 88–90, 237–238
guide, young Greek boy 81
Gully Beach 253, 270, 282, 283, 284, 323
Gully Ravine 15–19, 35, 44–45, 47, 102, 106, 192, 197–198, 207–208, 253, 257, 265, 269–270, 274, 285, *323*, *328*. *See also* Diversionary actions, 19 Dec 1915

INDEX

Gully Spur 253, 256, 274, *328*
Gurkhas 23

H

Hamilton, General Sir Ian 26, 265–266, 269
Harvey, Lieut. Frederick William 245, 304
Haylock, Group Captain E.F. 3
Heywood, Major W.D. 264–265, *286*
Hodgson, Bill 184
Holden, Lieut. Norman Victor *286*, 288, 304
Holden, Lt. G.G. *286*
Hornby, 2/Lt *286*
Howarth, Bill [Pte William?] 198, 249
Hughes, Major [C.E.] 307

I

identity discs 45–47, 89, 108, 167
Imbros 155–157, 234, 268, 270–271, 278, *322*
Imperial War Graves Commission 307
Ingham, L/Cpl Samuel 240, 297
Irish regiments 24, 45

J

Jimmy N—— 69–72, 131–132
Jones, 2/Lt *286*
Joyce, Capt. P.C. 287

K

Karoa, SS 273
Kemp, George. *See* Lord Rochdale
Kirte Dere 275
Krithia 250, 264, 323
Krithia Nullah 254, 276–279, *324–327*
 East Krithia Nullah 252
Krithia Road 278, 279, 323–327

Krithia Spur 251, 276, *324–327*
Kum Kale 99

L

Laffin, John viii, 225–226
Lancashire Landing. *See* W Beach
Lancashire men 3, 7, 21–22, 27, 114, 143, 171, 202
Leach, Capt. R.W. 268, *286*, 288
Lee, Brig. Gen. Noel 233, 261, 269
Lees, Major Roderick Livingstone 251, 264–265, *286*
Lees, Pte Harold 238, 298
Lord, Lieut. J.S. 268, *286*
Lord Rochdale 11, 29, 30, 37–40, 41–43, 109, 215–216, 231–234, *232*, 234, 246, 247, 248, 260, 264–265, 266, 267, 269, 276, 277, 278, 282, 283, *286*. *See also* Viscount Rochdale [son]

M

Macedonia 185–188, 221–222
machine-guns 52–54, 70, 113–115, 158–159, 256, 277, 285
 enemy 97, 266–268, 274, 275
 establishment 287
 sound of 36
MacLean, Capt., RFC 11, 186–188
Mahakian, Col. Carl K. 222, 224, 226–227
Marriatt, 2/Lt G.E. *286*
Marsden, Cpl Joseph 238, 298
Mathews, 2/Lt *286*
McPherson, Jock 22
medical
 dysentery 175–176
 field hospital 197
 hospital ships 191–192, 199, 200
 self-inflicted wounds 47–48
Mellor, Sgt John L. Taylor 114–115, 240, 298

INDEX 339

Menominee, SS 273
Mercer Road 276, 277, *324–328*
Molesworth, Lieut. Eric 268
Moorehead, Alan 189
Mudros 221, 284, 285, *322*
mules 34, 158
muleteers 34–35

N

Nelson, Sgt William Thomas 236, 299
Nil carborundum illegitimo 67
Nile, SS 219, 273
No-Man's-Land 55–56, 57, 67, 95, 113–114, 190, 200
Northern Barricade 283, *328*

O

O'Neill, Lieut. 286, 304

P

Parker, George 203
parsons. *See* chaplains
Pearson, Gunner Stanley vii, 149–153, 242–244
Platoon Sergeant 42, 122, 125, 136, 178, 199–200, 206, 210–212. *See also* No. 8 Platoon
puncing 25
punishment. *See* discipline, military

R

rations. *See* food and drink
Redmond, Lieut. W. 268, 289
Redoubt Cemetery 227, 306–307
Redoubt Line 279, *324–327*
reinforcements 116–118, 121, 287
River Clyde, SS 274
Robinson, Lieut. L. Maurice 268, 286, 288
Rochdale, Lord. *See* Lord Rochdale, Viscount Rochdale

Rochdale territorials. *See* 1/6th Lancs Fusiliers
Roe, 2/Lt 286
Royal Air Force 221–222, 223
  22 Balloon Company, 16 Wing 221
  balloon observers 185–187
Royal Flying Corps 48, 185–186, 221
Royal Naval Air Service 158, 309
Royal Naval Division 118, 282, 308–309, *324–325*
Royal Navy 33, 85, 93–94, 111–113, 307–308, 316
  minesweepers 98–99, 112, 155
  naval bombardment 37, 65, 99, 111, 112
  ships. *See Ermine, Karoa, Menominee, Nile, River Clyde, Uganda*
R——, Sgt 57, 62–63, 67, 236
RSMs 29, 61, 61–64, 64, 109

S

Savory, Sir Reginald 315
Scholes, L/Cpl William 240, 301
Scott, Capt. G. 264, 265, 267, 268, 286, 288
Scottish regiments 23, 54, 55, 308
Seddul Bahr 275, 323
Shaw, Major T.P. 3
Siddey-Botham 130
signals 19, 37, 192–193
  heliographs 94
Small Nullah 280, *324–327*
Smith, Lieut. Joshua Harold 244–245, *286*, 304
snipers 121, 148–150, 158–159, 243, 267
South African War, veterans of 31–32. *See also* Cpl I——; Griffiths, Capt. William H.; Lord Rochdale

Spafford, Capt. and Adjutant 264–265, *286*, 289, 304
Spain, old soldier 7–9
stand-to 80, 95–96
Stansfield, Cliff [Clifford?] 11, 124, 161, 241
Stanton-Jones, Richard 223, 227
Stirk, Johnny 121–123, 124, 240
Storey, 2/Lt C.B. *286*
storms 177–182, 254–255. *See also* winter
Stringer, Cpl 62
submarines 31, 33, 59
Sutcliffe, Tom 123–124, 241
Suvla 250, 254, 255, 268, 285, *322*
swimming 82

T

Tasker 140
tattoos 58–59
Taylor, Lieut. J. 268
Taylor, Lieut. T.R. 288, 304
Taylor, Lieut. W. 268
Templer, Sir Gerald 316
tobacco 20, 45, 52, 74–76, 135–141, 195
training 21, 216
  in Egypt 23, 29, 36, 65, 172–173, 219
  machine-guns 52–53
  Mudros 284
Trolley Ravine 282, 283, *328*
Turkish soldiers 65–66, 74, 97–98, 121, 148, 157–158, 178, 197, 204, 264
  fraternisation 179–182, 185, 187–188
  trenches 55
Turner, L/Cpl Frank 240, 302
Tweedale and Smalleys 123

U

Uganda, SS 273
Union Street 283, 284, *328*

V

V Beach 274, *323*
Vickers. *See* machine-guns
Vineyard, The. *See* Battle of 6/7 August – Battle of the Vineyard
Vineyard Trench (G12) 280–281
Viscount Rochdale 1, 3, 5, 43, 223–224. *See also* Lord Rochdale

W

Watkins, Doreen Mary 223, 227
W Beach 26–27, 220, 248, 253, 271, 274, 275, 285, *323*
Webster, Sgt Henry 236, 303
Whitfield, Frank [Francis?] 11, 73–74, 237
Wild, L/Cpl George 240, 303
winter 18, 34, 176, 189–192, 196–197. *See also* storms
winter quarters 256, 269, 285
Woodcock, Lt. Col. F.A. 284
Woolmer, Capt. E. *286*
Wyatt, 2/Lt G. *286*, 304

X

X Beach 282, *323*

Y

Y Beach *328*
yeomanry 24, 37, 231, 307–308
Yorkshiremen 143
Y Ravine 282–284, *328*

www.ingramcontent.com/pod-product-compliance
Lightning Source LLC
Chambersburg PA
CBHW022030290426
44109CB00014B/806